Samuel Morey and His Atmospheric Engine

Horst O. Hardenberg

SP-922

SAE HISTORICAL COMMITTEE

Prepared under the auspices of the
SAE Historical Committee

Published by
Society of Automotive Engineers, Inc.
400 Commonwealth Drive
Warrendale, PA 15096-0001
February 1992

ISBN 1-56091-240-5
Library of Congress Catalog Card Number 91-68251

Table of Contents

Abstract

From 1822 until about 1830, Samuel Morey from Orford, New Hampshire, developed an atmospheric internal combustion engine, which hitherto has been largely neglected in the history of technical science. This remarkable piece of machinery of progressive design burned a blend of alcohol and turpentine, mixed with air in a heated surface carburetor of which Morey is to be regarded as the inventor.

Despite several public demonstrations, he was unable to commercialize his patent rights, and the invention fell into oblivion without any influence on subsequent developments. Morey nevertheless has earned the merit of having built the first automobile in the United States - and worldwide the second - and also of having run it as the very first with liquid fuel, in Philadelphia.

As an essential part of the present paper, any document related to the origin and fate of the engine is reproduced in the appendix. Those that are only partially relevant to the subject are presented as partial extracts.

IT IS REMARKABLE that our motorized society, so far, has not brought forth any work which fully portrays the history of the development of the internal combustion engine from its earliest beginnings and with a worldwide view. Seen against this background, it is no wonder that Samuel Morey has remained a New England local hero (1,2)*, and even those interested in the history of technology have, if at all, only recently become acquainted with his successful efforts with regard to the atmospheric engine. The latter is primarily due to the unremitting efforts of Mrs. Alice Doan Hodgson who, having grown

---

* Numbers in parentheses designate references at end of paper.

up and lived in Morey's house in Orford, N.H. (3), has attempted to acquaint her fellow countrymen with his life and work (4). Credit is also due to C. Lyle Cummins, Jr. who made the Morey engine more widely known (5), without this leading to a fundamental historical study of the relevant material or to an appreciable echo in the literature.

Beyond studying Morey's internal combustion engine work, the present paper attempts to place in the spotlight a remarkable personality, who "having no pretensions to science, no chemical or philosophical apparatus, and little or no access to men of science, has spent much of his life in experiments" (6) and also accomplished inventive achievements which assure him an outstanding position in the history of technology.

That such a significant milestone in the development of the internal combustion engine has gone practically unnoticed, and the name of its inventor has remained nearly unknown, is a remarkable story in itself. And if in 1833 and 1834, Thomas P. Jones (1774-1848) (7), editor of the Journal of the Franklin Institute (8) and also contributor to the British Mechanics' Magazine (9), regarded the engine as well-known—"Brown's gas engine in England, and Morey's explosive engine in America, are familiar examples" [11]**—his conclusion was based on information acquired from 1828 to 1829 as Head of the U.S. Patent Office (10), and did not reflect the general state of knowledge. In Europe, for example, where Morey's invention had been published several times, the name of its inventor remained unmentioned (11-13), since the British patent on the engine [5] had been issued to a man named Erskine Hazard, referred to in a later chapter. The spelling "Herskine-Hazard," which is occasionally encountered originated from an advertising brochure of the Parisian firm Lefebvre of 1864 (14), one of the manufacturing plants of the direct-acting compression-less engine (15) invented in 1860 by Jean Joseph Etienne Lenoir (1822-1900). During this time period, the Old World regarded overseas technological developments merely in a superficial manner and with a certain contempt. Indeed, American patent specifications and technical literature were evaluated only in

** Numbers in brackets designate texts in the appendix.

exceptional cases. Thus, Morey's name could hardly become known. In later writings of technical historians, even the name Hazard appeared only as the inventor of a method for vaporizing fuels. This may be attributed to the title of his patent, "Method of Preparing Explosive Mixtures, and Employing Them as Motive Power" [5], which suggested more a mixture preparation process than an engine. Morey's engine, therefore, sank into complete oblivion in the last quarter of the 19th century—Hazard's patent being last mentioned in 1862 by Tresca (16), and not even considered in the comprehensive studies into the history of the internal combustion engine by Donkin (17), Clerk (18), and Güldner (19).

In order to save future researchers of the Morey engine the rather tedious procurement of papers and the deciphering of the unpublished manuscripts [1] to [3] and [9], all the principal sources are reproduced in the appendix. To facilitate their legibility, the spelling and word division have been modernized, and Morey's rather arbitrary punctuation—which sometimes even leads to a distortion of the meaning—has been modified in a few cases. In addition, it appeared necessary to include in the present study his technological environment, his other inventions, his personal circumstances, and—in view of non-American readers—their historical background.

Samuel Morey and His Age

HISTORICAL BACKGROUND

Although the establishment of colonies in the 17th century was based primarily on mercantilistic endeavors, the economic and social miseries of large parts of the European population, as well as religious dissidence, offered additional reason for emigrating. As a result, the population figure in the British possessions in America grew rapidly during the first two centuries of colonization (1620 - 2,000; 1700 - 275,000; 1776 - 2.5 million). The main representatives of the immigration to New England were Englishmen of the Reformed confession. Diligence, business acumen, strict morals and profound piety, but also narrow-mindedness in matters of faith, characterized the Puritan settlements which grew rapidly in the third and fourth decade of the 17th century. Their social order, initially theocratic and intolerant of dissenters, yielded only very slowly to more secular forms.

As early as 1639, Connecticut settlers conferred upon themselves the Fundamental Orders which greatly affected the constitutional development in the U.S., particularly with regard to the self-government of the communities. These were normally organized into townships—village settlements formally incorporated by the colonial parliament and vested with powers of self-government. Their size was usually 6 x 6 miles in area (9324 ha), and they encompassed, in addition to the individually allocated lots, wood and common. In the town meeting, to which only church members were admitted initially, all local matters were decided in direct democracy by the freemen. Many of the immigrants were not free at first, but had committed themselves to several years' service in order to pay off the costs of their passage. They met the manpower needs of the pioneer companies, and it was common practice for them to receive some land as their own property within five years and to be free.

While most of the British colonies initially belonged to private owners or trading companies vested with charters, growing administrative problems eventually resulted in their being placed under the authority of the Crown. The administration devolved on Governors, supported by councils, and was resisted by the settlers attempting to extend their rights. The Treaty of Paris, settling the French and Indian War (1754-1763) and making Britain the supreme colonial power in North America, was followed by an increasing animosity between the mother country and the settlers. Their westerly intentions were at odds with the British Empire Conception, and they opposed taxation by a parliament in which they were not represented. These conflicts led to armed confrontations, known by the questionable term American Revolution.

It began in 1776 with a number of colonies declaring their independence and forming an alliance with the aim of defending their freedom, and it culminated in the—second—Treaty of Paris (1783). The constitution of the United States, agreed upon in Philadelphia in 1787, was ratified the following year by the individual states. In 1800, the seat of the Federal Government became—after New York and Philadelphia—Washington, D.C. The conflicts did not end, however, until a further war had been fought (1812-1814), in the course of which British troops held Washington for two days (1814) and burned down the Capitol and other buildings.

BIOGRAPHICAL

Samuel Morey was born on October 23, 1762 in Hebron, Conn. (Fig. 1). He was the second of five sons and seven children of Israel Morey (1735-1809) and Martha Palmer (1733-1810) who had both been born in Hebron and had married in 1757. In the winter of 1765-66, the family moved to Orford, N.H., a few months after the first settlers. This place on the banks of the Connecticut River, the boundary between New Hampshire and Vermont, had already been chartered in 1761 and was named after the British Chancellor of the Exchequer, Robert Walpole (1676-1745), the first Earl of Orford, but settlement did not begin until 1765. In 1800, the population of the village was more than 1800; today it is nearly half as many.

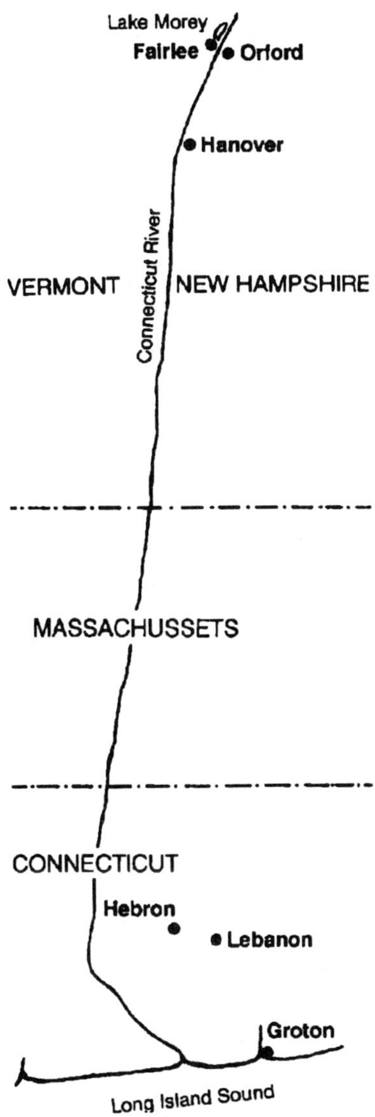

Fig. 1: Localities relating to Morey's biography

In Lebanon, Conn., Morey's father had been a pupil of Eleazar Wheelock (1711-1779) (7), the later founder of the Dartmouth College in Hanover, N.H., and he was considered to be highly educated. He established Orford's first store and inn, a ferry across the Connecticut River to Fairlee, Vt., a sawmill and two grainmills, and he ran a farm and a timber business. He made substantial donations to the community and performed numerous public duties. Thus, he supervised the public road works, and as a General of the New Hampshire State Militia, he headed the training of the recruits. Besides all this, he taught his children until Orford's first school was established in 1770.

Hardly anything is known of Samuel Morey's youth and education. His extraordinary experimental skills, which revealed themselves at a later date, suggest the conclusion that he had acquired not only excellent craftsmanship, but also a certain scientific knowledge, and it may well be that he owed this to Wheelock's library to which he had access since 1770 in Hanover. As all able-bodied men between the ages of 18 and 45, he served in the Militia, and as it was the custom in those days to address officers in private life with their military rank, he was later often referred to as "Captain" [9].

In 1785, he married Hannah Avery (1769-1822) who was born in Groton, Conn. Having built his own home in 1792 on the outskirts of the village, in 1799, he acquired a house which Orford's first minister had constructed on a plot of land purchased from Morey's father on the Ridge, a rise safely positioned above floodlevel. It was Orford's first two-story building, the front of which Morey completed in 1804 and which, in the first part of the 19th century, became part of what today is considered the "finest group of Federal-style houses in the U.S." (3).

Around 1790, the Moreys adopted a girl by the name of Mary Southworth, who later kept their house. The only child of their own was the daughter, Almira (1799-1830), who, in 1819, married judge Leonard Wilcox and received a wedding present of a house which her father had built. In its older part, dating from 1805, Morey's parents had lived, and it was linked to his own home by a covered footbridge.

The only Morey portrait was made only a few years after the invention of the atmospheric engine and two years before it was patented. It is part of a silhouette showing him with his daughter—not reproduced in Fig. 2—who died not long after having given birth to her sixth child.

In 1832, having already been widowed for ten years, Morey moved to the other side of the river to neighboring Fairlee, Vt. where he owned

Fig. 2: Samuel Morey in 1824

1500 acres (607 ha) of land—in those days largely pinewood around Fairlee Pond (which in 1890 was renamed Lake Morey by enactment). That the remarriage of his son-in-law led him to this move is unlikely, for the two were close friends and Morey later even provided in his last will for the children from his son-in-law's second marriage in the same measure as for his own grandchildren and his adopted daughter (21). As nothing is known of any age-related afflictions, and since he was no longer involved in business which might have necessitated a change of abode, it appears that he looked back dispirited and perhaps even angered at his many futile attempts to commercialize his engine and, therefore, abandoned his very well-equipped workshop.

Morey died at the age of 80 on April 17, 1843 at his home Bonny Vale in Fairlee, having taken his last dispositions—"weak in body, but of perfect and sound mind and memory" (21)—four weeks before. Although he had expended large sums on his developments, he left assets in a total value of $17,042, including four houses and a sawmill, as well as 3400 acres (1376 ha) of property on both sides of the Connecticut River worth $14,000, which were later divided into lots and sold in 1852. It is noteworthy and to a certain extent characteristic of Morey that he owned few books; in the list of his possesions compiled on the occasion of recording his last will, his "library" was estimated at only "$7.50" (21).

THE TECHNOLOGICAL ENVIRONMENT

The Industrial Revolution commenced in the second half of the 18th century in Britain and spread, with some time lag and lessened in its extent, to the U.S. Although mortality had considerably decreased and many country dwellers were moving into the industrial centers, the demand for manpower could hardly be satisfied. This led to a growing demand for mechanical energy, which could be met only partially by the slowly progressing introduction of steam power. Accordingly, there was a rapidly growing need for transport services, for which the network of available routes was completely inadequate.

European science had contributed hardly anything to the early development of industry, and such contributions were totally lacking in the U.S. This was due to the fact that universities—of which there were already 18 in the year 1800—were still purely educational establishments. The only scientific institution of the 18th century, the American Philosophical Society—founded in 1743 at the suggestion of Benjamin Franklin (1706-1790) in Philadelphia, Pa.—was not a research institute of European character such as, for example, the Royal Society of London and the Académie Royale des Sciences in Paris, founded in 1662 and 1666, respectively. The first American institution of this kind was the Franklin Institute of the State of Pennsylvania at Philadelphia, founded in 1824 (22). It became the model for further scientific establishments whose subjects of study were closely tied to the questions raised by industry, and which developed into promoters of applied science and technical development, while industrial companies initially did not pursue much technological research.

In the 18th century, there were already American journals which dealt with philosophical and scientific matters. Periodicals of primarily scientific and technical content, however, were not published before the first decades of the 19th century. Thus, there came into being in 1818 the American Journal of Science and Arts, edited by Benjamin Silliman (1779-1864) (7), Professor at the Yale College in New Haven, Conn., which became Morey's preferred journal for publishing his accounts (6,23). The American Mechanics' Magazine was founded by Thomas P. Jones in 1825, and the Franklin Journal was first printed in the same year. In 1826, Jones merged these publications to form the

Franklin Journal and American Mechanics' Magazine which was renamed in 1828 as the Journal of the Franklin Institute.

In view of Morey's rural environment and the limited possibilities existing up to the end of the first quarter of the 19th century for acquiring scientific or technological knowledge, his endeavors to obtain this with empirical means, which cannot be explained solely by the joy of experimenting, are astonishing. Equally impressive are his attempts to apply it for useful purposes, which were aimed not only at solving tasks of daily need, but at such distant goals as the powering of locomotives, boats, automobiles, and even of balloons [1].

HIS INVENTIVE ACHIEVEMENTS

Morey's inventions are likely to extend beyond the 20 patents issued to him in the course of 40 years (Table 1). He received 6 of the 306 U.S. patents granted up to and including 1800 (10). Yet, those for which he did not seek legal protection are, unfortunately, unknown.

1793	Turning a spit
1795	Mode of applying the force of steam
1796	Raising water by wind
1799	Power by applying water to steam
1799	Improvement to above
1800	Steam to raise water
1803	Steam engine
1812	Hot air power
1813	Fireplace
1815	Tide and water wheels
1815	Revolving steam engine
1817	Boiler
1817	Tide and current water wheel
1817	American water burner
1818	Improvement to above
1818	Treble pipe steam boiler
1819	Shooting with steam or gas
1819	Wire gauze for directing flame in stoves
1826	Gas or vapor engine
1833	Mode of decomposing and recomposing water

Table 1: Morey's U.S. patents

As the first patent for a steam-powered roast spit illustrates, his particular interest was centered on the utilization of steam power and eventually (1797) its use in boats. Morey cannot, however, be regarded as the inventor of the steamboat any more than Robert Fulton (1765-1815) (20). Steamboats had been operated in France as early as 1775 by the brothers Jaques Constantin (1742-1818) and Auguste Charles Périer (24), and in 1776 by Claude François Dorothée Jouffroy d'Abbans (c.1 751-1832) (24). And in the U.S., John Fitch (1743-1798) (7) had established in 1790 a regular passenger service on the Delaware River operating from Philadelphia with his steamboat "Thornton," named after a man referred to in a later chapter.

The combination of steam engine and paddle wheel was not new, and had been described by Jonathan Hulls (1699-?) (25) in a British patent of 1736 (26) and a book (27). Morey's two paddle wheels arranged on the sides of the boat also were not original. This concept had been anticipated in drawings by Leonardo da Vinci (1452-1519) and in the practical application by Denis Papin (1647-1712?). What was novel was Morey's combination of such wheels with a steam engine. It was regrettable that a patent issued to Fulton in 1811 prevented Morey from using his own invention, not to speak of exploiting it commercially. Fulton could not be blamed for this, however, as Morey is said to have—"Blast his belly! He stole my patent!" (4)—since the resulting patent award was due to the new U.S. patent law effective of 1793, which, in line with British law, had abolished the examination for novelty, reintroduced in 1836, and merely provided for the registration of inventions (10). Morey's most significant, yet astonishingly less heeded, invention was his internal combustion engine. Based on the patent grants [3] and [5] and his own publications (23,28), it is usually dated to 1826 (5). It appears, however, to have already originated prior to 1822, probably from experience described in his account "On Heat and Light" (6) published in 1820. This engine was in many design features far in advance of the technological possibilities of the time and formed the climax of Morey's creativity. After this point, he applied for a patent for only one further—insignificant—invention.

Morey's Atmospheric Engine

THE ENGINE DESCRIPTIONS

Compared with the often scarce information on other inventions from the early childhood of the internal combustion engine, the available material regarding Morey's development work can be considered as exceptionally good. Morey himself left behind four descriptions of his engine, written between 1824 and 1826; an even older text must, regrettably, be deemed as lost. In addition, there are reports from two credible eyewitnesses capable of rendering a professional judgment.

The documents involved are the following seven, some of which require clarification in greater detail, as is done below:

- Morey's manuscript of October 1824 [1].
- Morey's undated specification draft [2].
- Morey's U.S. patent of April 1,1826 [3].
- Hazard's British patent of August 12,1826 [5].
- Morey's letter to Prof. Silliman of 1826 [6], which was, curiously, published in the Franklin Journal and American Mechanics' Magazine (28) some months earlier than in the American Journal of Science and Arts (23), edited by Silliman himself.
- Eyewitness report on Morey's engine exhibition in New York [12] (30) by Thomas Ewbank.
- Eyewitness report on Morey's engine demonstrations in Philadelphia [13] (31) by George Escol Sellers.

Morey's manuscript of October 1824 [1] bears an inscription in his own hand on the cover page: "Copies of papers sent to Secretary of State for a patent New Power. June 1822, a mistake. Then papers prepared for Publication Oct. 1824, but not sent. They are copied."

The dating "Octr 5," as is found in (4), is based on an erroneous reading of the superscript r as a 5. It was the administrative practice in those days that caused Morey to apply for a patent from the State Department, i.e., from the Ministry for Foreign Affairs. The Patent Office, when founded in 1802, was placed under the State Department's authority on the justification that patents were to be seen from an international viewpoint—in 1849, it came under the jurisdiction of the Department of the Interior (10). Morey's patent application of 1822 should not be equated with the undated specification draft [2] which certainly originated at a later date than the manuscript [1], as comparisons made in later chapters clearly reveal, and is likely to have been composed in 1825.

It should be noted that up to 1849, U.S. patents were signed by the President, the Secretary of State, and the General Attorney. Thus, Morey's engine patent [3] was signed by the sixth U.S. President, John Quincy Adams (1767-1848), who held office from 1825 to 1829. The specification of the invention bears not only Morey's signature, but also the attestation of two hitherto unknown witnesses: "Alex McIntire William Thornton." Their identification leads to the conclusion that Morey is likely to have gone to Washington on the occasion of his patent application, a destination hitherto unknown in his biography. Alexander McIntire was an employee of the remarkably small four person Patent Office (33). The second witness was Dr. William Thornton (1759-1828) (7), based on the comparison of the attestation on Morey's patent with other of Thornton's signatures (34). Originally a physician, he had as an architect designed the first U.S. Capitol which was burnt down in 1814. He was appointed "Keeper" of the Patent Office in 1802; he called himself "Superintendent," a title not legalized until 1830, two years after he had left the office. He is also said to have persuaded the British in 1814 to spare the Patent Office for "what would be useful to all mankind." Nevertheless, all the files relating to Morey's inventions were lost in the Patent Office fire of 1836. It is conceivable that Morey had been in contact with Thornton at an earlier date as a result of his steamboat activities, since the latter had likewise concerned himself with steamboats, together with John Fitch, and had written a book on their origin (35).

That the subject of Hazard's patent [5] was Morey's invention cannot be seriously doubted. As early as April 1826, the American technical world had been informed (29): "Samuel Morey... has obtained a patent for a vapor engine. ... A gentleman has gone to England, for the purpose of obtaining a patent in that country" [4]. Oddly enough, British law permitted a patent to be issued to an applicant resident in Britain in place of the real inventor. This applicant and certainly Morey's authorized representative was the American engineer Erskine Hazard of whom little is known, except that he had previously been involved in the floating of coal on the Lehigh and Delaware Rivers (32). Hazard himself did not claim that the invention described in his patent was his own, but expressly stated that it was "partly communicated by a certain foreigner residing abroad, and additions made by myself" [5]. Of the small number of Morey's test parts remaining—preserved in the Vermont Historical Society Museum in Montpelier, Vt. and in the Morey House in Orford, N.H.—there are unfortunately none from his engine. Photographs of the engine (5,54) originate from two reproductions. One is in the collection of the Smithsonian Institution in Washington. The second was on display at first in the New York Museum of Science and Industry and later in the Long Island Automotive Museum in Southampton, N.Y., and has meanwhile passed into private possession (Peter Levesque, Hartland, Vt.).

The significance of Morey's engine descriptions lies in the fact that they permit us to reproduce the development of the engine as far as its design features are concerned. They do not reveal, however, how he arrived at the basic principle of the atmospheric internal combustion engine. Namely, there is no hint of how he discovered the method of producing hot combustion gases in a cylinder, which, when cooling down, form a vacuum which—as was common from the atmospheric steam engine—causes the atmospheric pressure to move a piston, thereby creating a mechanical force which can be used in a multitude of ways.

STATE OF ENGINE TECHNOLOGY

In studying the question of whether the principle of the atmospheric engine was the result of Morey's own reflections or, perhaps, was due to outside stimulation, and to what extent he attempted to separate himself from other engine developments, the following fact is encountered.

In his U.S. patent [3], he laid claim primarily to the operating principle of the engine, and less to the mixture preparation method:

- The mixture formation outside the cylinder: "mixing... air and vapor or gas... previous to their entering into the exploding chamber."
- The production of a combustion gas volume greatly enlarged over that of the mixture, in order to obtain a powerful vacuum: "making use of... that explosive all; and thus producing by explosion a much larger or more extensive vacuum than the space the prepared air previously occupied."
- The provision of a novel prime mover and its practical application: "furnishing in a new way and on new principles a first moving power, and the application of that power to useful mechanical operations."

By contrast, the British patent [5] only stated:

- The formation of stoichiometric mixtures outside the cylinder: "the method of mixing... with the inflammable materials, previous to their entering the exploding chamber, such portions of... air as they would combine with in combustion."
- The separation of mixture and air in the cylinder, as well as the method of expelling the latter and keeping it from being involved in the combustion: "keeping the explosive mixture separate from the atmospheric air in the cylinder and driving out the latter without making use of it in the combustion." - The particular design of the exhaust valve: "the peculiar construction of the flexible valve."
- The mixture formation far below the boiling temperature of the fuel: "preparing the explosive mixtures at a temperature much below that of the boiling point!"

It is only possible to guess as to the reason for the essentially different claims of the British patent [5], which imply a restriction to the mixture preparation method and certain engine design characteristics. Such restrictions could not have been due to objections of the patent authority, since British law did not provide for any examination for novelty prior to 1883 (36). Perhaps Hazard's "additions made by myself" [5] consisted mainly in the restriction of the claims and the rewording of the title which in the case of the U.S. patent [3] referred solely to a "Gas or Vapor Engine," whereas the British patent put the emphasis on the mixture formation: "Method of Preparing Explosive Mixtures, and Employing them as Motive Power." Hazard may have made these revisions after learning on his arrival in England that, shortly before, in May 1826, an engine based on the same principle was shown in a public presentation installed in a vehicle (37). This had been built by Samuel Brown (17??-1849?) and was based on a British patent issued in April 1826 (38), which contained improvements compared to a patent dating from 1823 (39) and a U.S. patent of identical content of 1824 (40).

Morey himself must have been aware of other engine developments at an even earlier date, as can be derived from the specification draft of 1825 [2] in which he stated that he named his invention "vapor engine, to distinguish it from the... gas engine." As the atmospheric working principle was already an essential element of his engine of 1822 [1], the question arises how much Morey could have known about the contemporary state of engine technology.

It can be assumed that Morey had never heard of the atmospheric gunpowder engines of the Renaissance or the Baroque period (41), linked with names such as Leonardo da Vinci, Christian Huygens (1629-1695), Jean de Hautefeuille (1647-1724), and Denis Papin. Although he may well have learned of the engine patented in 1794 by Robert Street (42) as much as of the one proposed in 1801 by Philippe Lebon d'Humbersin (1769-1804) (43), these could not have served as models, since they were direct-acting. The same applies to the machines patented to the brothers Joseph Claude (1763-1828) and Joseph Nicéphore Niepce (1765-1833) in 1806 (44,45) and 1817 (46), and also to the gunpowder engine built in 1807 by George Cayley (1773-1857) (47).

That Morey could have heard of Samuel Brown's engine of 1823 (39) only after having invented his "New Power. June 1822" leaves only three atmospheric engines to which he could have had recourse, yet it is extremely unlikely in all cases that he had knowledge of them. These involve the engine patented by Isaac de Rivaz (1752-1829) (49) in 1807 (50,51), the engine designed in 1820 by William Cecil (1792-1882) (52), and an unnamed inventor's engine which Thomas P. Jones had examined in 1814 [7] (53), without ever publishing its design features or the test results obtained.

Even if outside stimulus cannot be totally ruled out in view of the multitude of contemporary engine development efforts, many of which remained unknown since they were unsuccessful, until evidence to the contrary is presented, Morey has to be regarded as an independent re-inventor of the working principle of the atmospheric internal combustion engine and as the creator of a machine for continually performing the respective working cycle

THE EVAPORATOR

In all of Morey's papers, basically the same evaporator design is described, namely a rectangular box made of tin, brass, copper or other sheet metal, for which he indicated very different outer dimensions. According to [1], length and width were about 30 cm and the height 18 cm or more ("about one foot diameter; and seven or more inches deep"). According to [2], it should "not have more area than three times that of the cylinder of the engine," yet it is not clear which areas are meant in the case of the evaporator and cylinder, respectively. In [3] and [5], no figures are given in this regard, and according to [6], the length should be about 36 cm, the width between 10 and 13 cm and the height 18 cm ("four or five inches wide, and about fourteen long, and seven deep").

As Fig. 3 from Hazard's patent [5] shows, this box was "divided horizontally into compartments by partitions which extend alternately from one end of the box to within a short distance of the other end." Here again, the data differ: In [1], a spacing of the partitions of between 51 and 76 mm is stated ("once in two or three inches"), with "the space between the upper division and cover being about" 25 mm ("one inch") "deep." In [2], no such figures are given at all. In [3], 4 or 5 horizontal partitions corresponding to "five or six compartments"

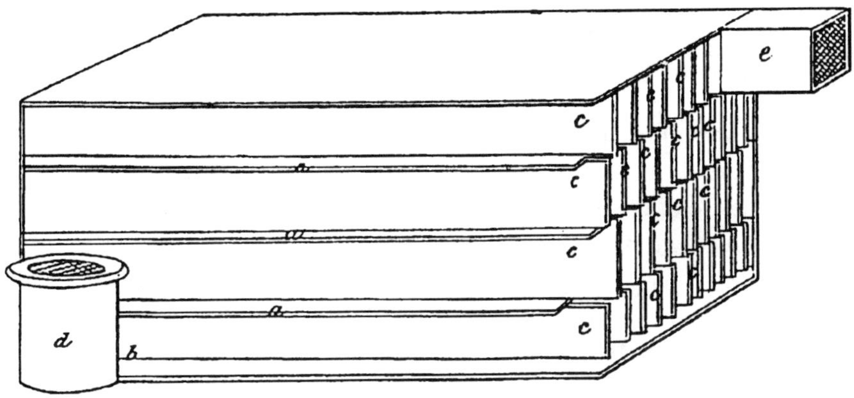

Fig. 3: Evaporator (from Hazard's patent drawing)

are mentioned, in [5] only 4 compartments, and in [6] 4 or 5 of those. The figure contains a drawing error to the extent that the bottommost and uppermost horizontal partitions do end before the righthand outer wall of the box. The middle partition, however, extends up to the lefthand outer wall, which is why the box would not permit any passage to the flow.

The evaporator was also to be divided by vertical partitions, for which Morey again gave differing figures: In accordance with [1], they were to have a spacing of 13 mm or more ("half an inch or more") and be "in a zigzag or spiral form." According to [2], they should form straight spaces "communicating at alternate ends" and were to have "a range of small holes in them." In accordance with [3], [5], and [6], the partitions should form straight spaces about 13 mm wide ("about half inch").

The box was to have two openings. According to [6], each "a short tube" of "about two inches in diameter ... one for the admission of air and other materials into the lower apartment, and another for letting out the... prepared air from the upper one." Both openings were to be "covered with fine wire gauze, to prevent explosion" [6], namely "of the fineness directed Sir Humphrey Davy" [2] (1778-1829), the inventor of the mine safety lamp. In [1], use of "a safety valve of any size which need not be loaded at all if it weigh but two or three ounces" has also been thought of. Why Morey, in [2], only assumed "the loss of its top" in the event of an explosion in the evaporator, is not clear, however, since he made no mention of any predetermined breaking point in the box, which would be necessary in this case.

The air was to be "driven in by a bellows" [2] "on one side" [1] of the box. In addition, Morey had made provision for controlling the temperature of the mixture and thus its air/fuel ratio by means of bypassing the evaporator, using "a second pipe leading from the bellows to the upper part of the evaporator to admit a stream of cooler air" [2], or by leading air "into a space in the box, above the lower one, or that containing the liquors" [1]. That in [3], [5], and [6], no further mention was made of this mixture dilution leads to the conclusion that it had proven to be unnecessary, providing the evaporator temperature was carefully controlled.

For this purpose, Morey considered any form of evaporator heating as being suitable, namely "a lamp, or other small quantity of fire, or... the sun's rays, or other source of heat, as fermenting grains, etc." [2], "or other heating substance" [3]. The figures stated for the temperature to be achieved are, however, curious: Whereas in [2], mention is made of a heating to 27°C ("80° of Fahrenheit"), according to [1], [3], [5], and [6], the box was to be warmed to "about blood heat," which is contradictory to the extent that the figure ascribed to the temperature of the human body—in accordance with a proposal of Isaac Newton (1642-1727)—is 100°F (37,8°C).

Morey regarded an evaporator of about 36 cm in length, 11 cm in width and 18 cm in height ("fourteen inches long and four and a half wide, and seven inches deep") as "sufficient... to prepare air fast enough, to furnish from fifty to one hundred charges per minute, for a cylinder" of 18 to 20 cm diameter ("seven or eight inches") and 61 cm stroke ("two foot") [6]. This figure was based on the observation that "only 1/8 of the cylinder is required to be filled with the explosive mixture, and from experiments it ought to be less than 1/12" [1], and the remainder with air.

Morey's evaporator represented a heated surface carburetor in the bottommost compartment, while the upper compartments were intended only to provide for the best possible mixing of fuel and air by deflecting the flow the maximum possible number of times.

THE FUELS

Compared to all other contemporary internal combustion engine work, it is particularly remarkable that Morey, right from the beginning, primarily considered the use of liquid fuels. He also limited such fuels to only those which can be converted, together with air, into inflammable mixtures "at a temperature much below that of the boiling point" [5]. Although he did make mention of the possibility of employing hydrogen, this was done only in view of the potential use of his engine "forgiving ... direction and velocity to balloons" [1]. Similarly, when he considered natural gas (methane CH_4)—what he called "hydrogen gas," from undefined sources ("springs which are constantly giving off hydrogen gas") [3]—as being useful, he did this primarily to illustrate the versatility of his engine and its independence from the type of fuel used. That he did not perform any experiments with gaseous fuels, at least nothing is known of such, may have been due to their lack of availability in his home village. Even for coal gas which was just beginning to gain a foothold, there was no supply network, even in the industrial centers, and the first U.S. gas companies had just been founded—1816 in Baltimore and 1823 in New York (5). It may be that Morey, intuitively perhaps, but correctly, regarded liquid fuels as being more universal in application and equally well suited for stationary and mobile use, even though he did not expressly state this.

Although he stated all "materials, conveniently capable of evaporation, and the vapor of which is inflammable when mixed with atmospheric air" [3] as being suitable for his engine, the fuels which he used himself were solely alcohol and turpentine. Whether the former was the respective azeotrope, namely "highproof spirit" [3], or a low-percent drinking alcohol ("whiskey") "or other cheap spirit" [2], was apparently of no concern to him. Chemically, it was ethyl alcohol (ethanol C_2H_5OH) diluted with more or less water, having a molecular weight of 46.06, a density of 0.785 g/cm^3, and a boiling temperature of 78°C. Turpentine which he mostly used as an additive to alcohol, is the low-boiling portion of the pine oil distilled from chopped pinewood (pinus palustris). It is a thin-bodied, essential oil with densities between 0.855 and 0.972 g/cm^3 and a boiling range from 155 to 166°C, and consists mainly of cyclic terpenes of the form $C_{10}H_{16}$ with a molecular weight of 136.24. In terms of quantity, these are particularly α-pinenes with a density of 0.859 g/cm^3 and a boiling point of 156°C and a smaller portion of B-pinenes with densities between 0.865 and 0.869 g/cm^3 and a boiling range between 164 and 166°C.

The various fuel components result in the vapor pressures listed in Table 2 at the temperatures stated by Morey of 70, 80 and 90°F and of "blood heat" (100°F). The values for turpentine have been calculated from the vapor pressures of the components, assuming a molar composition of 80% α-pinenes and 20%, β-pinenes. Depending on fuel composition and temperature, these result in considerably differing air/fuel ratios which, however, must not vary greatly from the stoichiometric air demand of the respective fuel.

The latter is largely dependent on the fuel composition insofar as the air demand for stoichiometric combustion substantially differs for ethanol (9.00 g air/g ethanol) and for turpentine which was assumed to be a pure pinene mixture (14.2 g air/g pinene).

Temperature		Water	Ethanol	Turpentine
°F	°C	mmHg	mmHg	mmHg
70	21.1	18.8	44.1	3.9
80	26.7	26.2	59.9	5.3
90	32.2	36.1	80.4	7.1
100	37.8	49.2	107.1	9.6

Table 2: Vapor pressures of the fuel components

Table 3 indicates the resulting relative air/fuel ratios (the Figure 1 represents stoichiometric combustion) which would be obtained with various fuels, if the corresponding state of equilibrium of the vapor pressures was reached in the evaporator. As this cannot be achieved, but is only more or less approximated depending on flow rate, higher air/fuel ratios, i.e., markedly leaner mixtures, have to be assumed for practical operation of the Morey engine. In addition to pure ethanol and turpentine, as well as to a mixture of equal volumetric portions of both materials, consideration was given to a mixture consisting of equal

Temperature °F	Ethanol 100%	Turpentine 100%	Ethanol + Turpentine 50 + 50vol.-%	Ethanol + Turpentine + Water 40 + 40 + 20vol.-%
70	1.13	2.90	1.10	2.45
80	0.82	2.13	0.80	1.76
90	0.59	1.59	0.58	1.30
100	0.43	1.17	0.43	0.96

Table 3: Fuel composition, temperature, and relative air/fuel ratio

portions (40%) of ethanol and turpentine and of 20% water. This latter percentage approximately equates to the solubility limit of water, as turpentine is miscible with ethanol in any ratio, but insoluble in water.

As can be derived from Table 3, the ethanol-air mixture produced with Morey's evaporator is likely to have had an approximately stoichiometric air/fuel ratio (corresponding to a relative air/fuel ratio of 1) at an evaporator temperature of about 75°F, and nearly the same holds true for the mixture of alcohol and turpentine. In contrast, engine operation with pure turpentine, as is mentioned in [7] and [12], is likely to have been possible only at a temperature "about blood heat" (100°F). As Table 3 also shows, the admixed water had the effect of leaning the mixture and of, undesirably, requiring a higher evaporator temperature.

HIS MIXTURE FORMATION EXPERIENCE

It is typical of Morey's modesty that he ascribed the invention of his engine to mere chance ("accident has given to the mechanical world a new power") [1]. In so doing, he was thinking of his accidental discovery "that the vapor of water and that of spirit of turpentine, when mixed with a very great proportion of atmospheric air, were highly explosive" [6], and he assumed that this was "new to chemists of the highest reputation" [2]. This discovery appears to have been made after he had completed the experiments described in his account "On Heat and Light" (6), published in 1820. This can be concluded from the fact that therein, no mention is made of the inflammability of the aforementioned turpentine-air mixture, nor of a device for its preparation. On the other hand, he must have acquired this experience prior to the summer 1822, for it was already an essential element of his "New Power. June 1822."

How thoroughly Morey attempted to clarify the laws of fuel evaporation and mixture preparation with the very simple means at his disposal is illustrated by his various observations, namely:

- that a fuel with a lower percentage of alcohol requires a higher evaporator temperature: "the less proportion of alcohol is used, something higher temperature is required" [3].
- that turpentine hardly evaporates at a temperature of 21° C and thus scarcely contributes to engine power output: "at about 70° the spirit turpentine refuses to be taken up by the air or at least in sufficient quantities to give energy to the explosion" [1].
- that this observation is fundamentally valid for all fuel components, and that the thus produced—too lean—mixtures burn only slowly and with low effectiveness: "at a very little lower temperature 70° they do not evaporate freely enough to give energy to the explosion, or to explode quickly and effectually" [2].
- that too-rich mixtures produced at too-high evaporator temperatures burn only sluggishly: "so on the contrary at a higher temperature, the vapor rather burns than explodes" [2].
- that in this case, more air must be blown through the evaporator: "if the temperature is considerably increased above blood heat, it will require an increase in the quantity of air" [1].
- that with higher flow velocities in the evaporator, more mixture can be prepared at the same temperature: "it appears that the more rapid the current of air is made to pass over the surface of the liquor, the more rapidly it takes up vapor at the same temperature" [3].
- that the longer the distance which the air travels along the surface of the liquid fuel, and the greater the number of deflections, the lower the needed temperature: "the more turns the air makes in contact with the liquors in the box, the lower temperature will be required to prepare it" [3].
- that more fuel-air mixture can be produced at a more uniform air flow velocity, as is achieved when the evaporator is utilized on a multicylinder engine: "the box being in use too

only about 1/4 part of the time, it of course would supply in the same way, four such cylinders if the air were blown or drawn constantly through it" [3].

- that in this case, more heat needs to be supplied to the evaporator: "to keep up the temperature of the box in that case would require the application of more heat" [3].
- that the addition of ethyl ether (C_2H_5-O-C_2H_5), in a concentration adapted to the ambient temperature, would permit operation without evaporator heating: "with the addition of a due proportion of ether, according to the temperature of the weather, it is likely the engine will work any time with only a lamp to fire the charges" [3].

In summary, Morey had realized that his engine could not be operated with any fuel-air mixture, and that there are flammability limits with both rich and lean mixtures. This explains why he cared for the control of the air/fuel ratio by means of air bypassing the evaporator with which "we can regulate the mixture as we please" [1]. Also, he had experienced that the air/fuel ratio to be obtained is dependent on the fuel components.

What is not correct, however, is his conjecture that the mixture composition is fairly independent on the portions of the fuel components ("it does not appear essential in what proportions to each other the liquors are put into the evaporator, as the laws of elective attraction among the particles appears to be undisturbed by the quantities") [2]. He thus believed that equal parts of both components evaporated from any mixtures of water and turpentine ("when water and spirit of turpentine are put in together, in equal quantities, or unequal, the air takes up equal parts of each") [2], which statement is dubious due to the fact that turpentine is insoluble in water. In the case of water-alcohol-turpentine mixtures, he proceeded from the belief that the fuel components evaporated in the ratio 2/2/1 ("if alcohol be added to them, the air takes up two parts of the spirit, two of water and one of the turpentine") [2] and that this ratio was 8/1 in the case of alcohol-turpentine mixtures ("if alcohol and spirits of turpentine are put in without water, the air takes up nearly eight parts of the former to one of the latter") [2]. Interestingly enough, such ratio figures are to be found only in the manuscripts [1] and [2], while later statements are clearly moderated insofar as they term the fuel composition as being "not very material" [6]. This change in opinion may have been based on new experience.

CYLINDER, PISTON AND ENGINE DESIGN

The cylinder was to be made of "sheet brass or copper, and even of tin plates" [1]. Which material actually was used is uncertain, although it can be assumed on the basis of experiments devised in [1] ("a cylinder ... made of tin plates, I have repeatedly filled ... with the ... mixture") that it was tinplate, at least in the early days of the development. Equally uncertain are the cylinder dimensions. Morey himself mentioned only one cylinder of almost 30 cm diameter ("nearly one foot") and 46 cm stroke ("eighteen inch") [1], while corresponding figures of 1 and 3 feet in [2] as well as 8 inch and 2 feet in [3] and [6] are only calculation examples. According to the eyewitness report [12], the engine demonstrated in New York had a cylinder diameter of about 23 cm ("about nine inches") and, in accordance with the report of the engine demonstration in Philadelphia [13], of about 25 cm ("about 10 inches"). No data are available regarding the piston stroke in either case, and it is also questionable if the stroke/bore ratio of 2.68 as shown in Hazard's patent drawing [5] correctly reflects the true dimensions of the engine.

The gradual development of the engine can be clearly seen from the example of the number of cylinders considered to be practical. Whereas the oldest manuscript [1] speaks only of one cylinder, in [2], the use of several cylinders is already mentioned as useful with regard to the uniformity of the power output ("by employing a third cylinder it will be a continued unremitting power ... four cylinders ... would have the same consequence more powerfully"). This development is also one of the indications that the undated specification draft [2] originated later than the manuscript [1], and that it should not be equated with the "papers ... for a patent New Power. June 1822." In his later texts [3], [5], and [6], Morey regarded a second cylinder operating at a crank angle offset by 180° as an essential element of his engine, which was actually used as the eyewitness reports [12] and [13] verify.

The information available on the piston design is extremely sparse. In [2], for instance, only scanty reference is made that the piston crown, called "bottom," was made of "sheet metal," a material which Morey seems to have retained in the form of "sheet copper," as is confirmed by the eyewitness report [12]. Whereas no respective statements are made in [1] and [2], the patent specifications [3] and [5] speak of two different types of construction, namely of "plunger or piston." That these are not different terms for the same part is evident from Morey's remark in [6] that he had "substituted a plunger, in lieu of the piston," as well as from the eyewitness report [12] where a similar statement is made ("the pistons were made ... in the form of plungers").

Conclusions regarding the piston design can be made from the sectional drawing of the British patent [5], part of which is reproduced in Fig. 4. This is, however, subject to the proviso that the pistons actually used in Morey's engines may have looked quite different, particularly with regard to the design of the piston seals, of which he considered many different types.

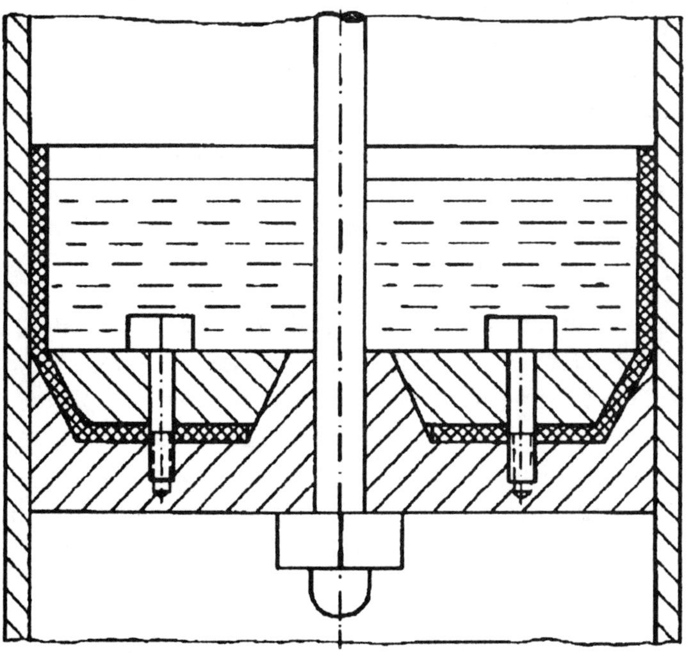

Fig. 4: Piston design [5]

According to Morey's own descriptions, leather was used for this purpose in all cases. A comparison of the descriptions given in [1], [2], and [3] reveals how the seal design progressively developed. Initially, a metal ring surrounding the piston of approximately cylinder diameter and a height between 5 and 15 cm was applied, which was to hold two or more leather cylinders projecting a few inches beyond the metal ring, which should be moderately pressed against the cylinder wall by springs or a light packing ("I... enclose it in a hoop of thin metal, that nearly fits the cylinder, from two or three to five or six inches deep; within that hoop, I put one within another, two or more cylinders of leather, made tight to the piston at bottom, and projecting some inches above the hoop. This cylinder of leather is pressed moderately against the sides of the cylinder by springs, or by some light elastic stuffing") [1]. In the subsequent and only slightly refined version, the leather should be kept moist by water or oil filled onto the piston ("make the bottom of metal, and fit a ring or turn it upward like a dish, so as nearly to fit the cylinder; filling the space nearly with a wooden piece firmly securing it to the sheet metal bottom of the piston, surround this with one, two or more rings of leather which will rise above the ring of metal enough to apply closely to the bore of the cylinder: Place around the piston rod a spiral spring or springs, or other elastic substance gently to press the leathers to the bore, and let water or oil be put on the piston to keep the leather moist") [2]. The seal design culminated in a single leather cylinder ("I put round the piston a hoop of thin to five or six inches deep, and a cylinder of leather with in that projects above two or more inches, gently pressed to the sides of the cylinder") [3].

This latter seal design in no way corresponds to that reproduced in Fig. 4, which is characterized by a leather cup seal held by a ring screwed to the piston. Interestingly enough, the liquid filling of the piston—as Huygens had used 250 years earlier in his gunpowder engine to keep the piston seal moist and press it against the cylinder wall (41)—is mentioned in [2] ("let water or oil be put on the piston to keep the leather moist"), but not in the British patent [5], although the patent drawing clearly shows it.

It is unknown how the seal mentioned in [6] ("I ... attach the stuffing to the top of the cylinder; it works well, and is more convenient") may

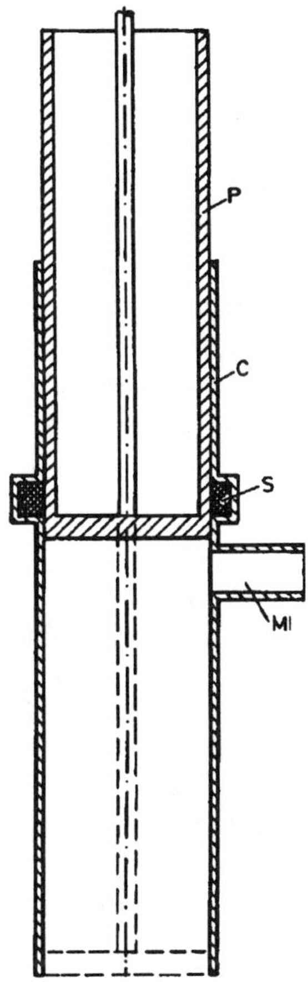

Fig. 5: Presumed plunger design

have looked. This appears to have been the design used during the engine demonstration in New York. According to the eyewitness report [12], the piston seal was "a strip of oiled listing or cloth tied round the upper ends of the cylinders." However, this statement contradicts Morey's that all his seals were of leather, while [12] talks of textile material ("listing or cloth"). Fig. 5 shows how the "plunger, in lieu of the piston" [6] or the "pistons ... in the form of plungers" [12] and the seal at "the top of the cylinder" [6] might have looked. In contrast to the short, skirtless piston with its moved seal as shown in Fig. 4, the plunger is likely to have had a skirt of such a large length as to permit the piston to be sealed equally well in any position between its bottom dead center and top dead center by a packing S fixed to the cylinder C. Such a design has the benefit of keeping the mixture inlet MI closed all along the downward stroke of the piston P.

It can be assumed that Morey's engines differed from the drawing of the British patent [5] not only in the piston and seal design, but also in further essential points. Those differences can partially be explained by the fact that the drawing mentioned did not originate from Morey himself, but first in England, and it is very likely to have been prepared by a professional engineer, perhaps from Hazard himself. This presumption is supported by the fact that Morey at the same time declared himself unable to enclose one in [6] ("a drawing ought to accompany this paper, but at present, it is out of my power to furnish one"). It is worth mentioning that Morey's remaining manuscripts do not contain a single sketch, from which it may be concluded that all his work was performed without any graphic basis.

As comprehensively as Morey described many a component of his engine, he barely expressed himself as regards its crank assembly. The cause for this might have been that he judged its design, which today strikes us as curious, to be sufficiently familiar to the expert and therefore not in need of any explanation. For the reason of simplification, Fig. 6 schematically represents only a single-cylinder crank assembly, whereas Hazard's patent drawing in [5] represents the actually used two-cylinder version.

The piston P in the cylinder C was connected by the piston rod PR to the crosshead CH. Its outer ends slid in two vertical guides and were linked by two parallel connecting rods CA to the crankshaft CS, which was mounted in a frame not shown in Fig. 6. This arrangement keeps normally-acting forces from the cylinder walls. In the two-cylinder engine, the cranks at both ends of the crankshaft were supported by bearings on one side only, as is the one on the flywheel side of the engine in the figure. Between the cylinders, a large gearwheel LW was mounted on the crankshaft, which meshed with a smaller gearwheel SW while reversing the direction of rotation and increasing the rotational speed—in the ratio 1 to 4.5 on the basis of a count of the number of teeth in Hazard's patent drawing—and transmitted the power to a fly-wheel FW by means of a driveshaft DS parallel to the crankshaft CS.

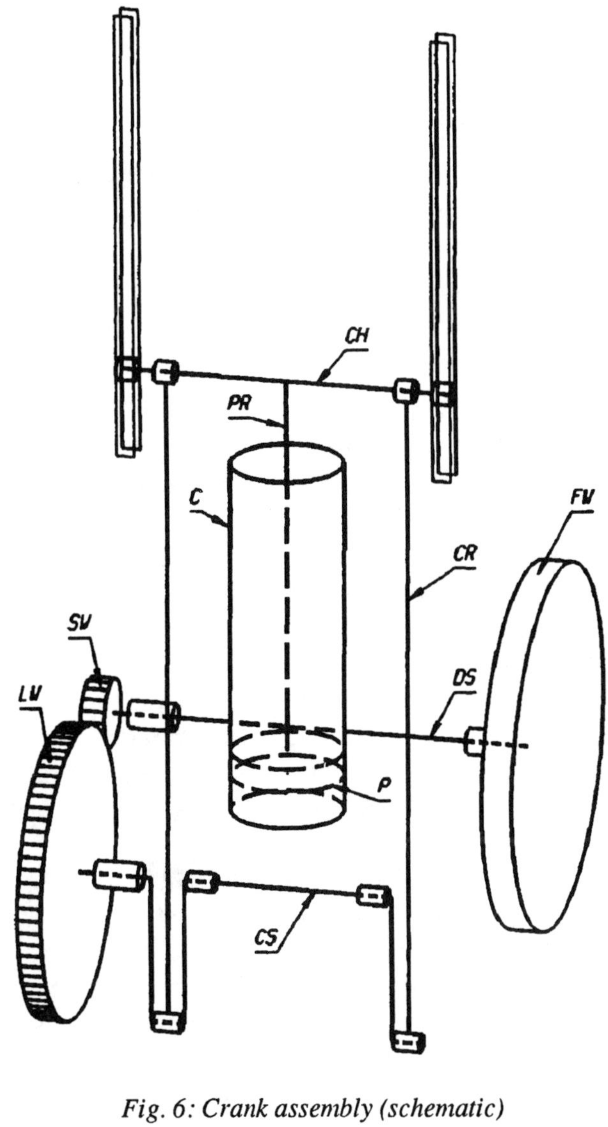

Fig. 6: Crank assembly (schematic)

Over the entire development period, Morey retained this design with the cylinder directed downwards, filling being performed during the upward stroke of the piston and work on the downward stroke, as is revealed by the eyewitness report [13] ("on the upstroke ... atmospheric air and vapor were drawn in ... and the piston descended by atmospheric pressure into the vacuum").

VALVES, IGNITION AND CONTROL

The engines of 1824 and 1825 were characterized by the use of the same valve for filling the cylinder with air and for evacuating it. It was located at the bottom end of the cylinder and should have 1/4 to 1/2 of the bore diameter and open outwards ("a cylinder ... to the bottom of which we attach a valve to open downwards, of about from 1/4 to 1/2 the diameter of the bore of the cylinder") [1]. It should have a particular type of seat and a low flow resistance and be rendered airtight with leather or other material ("the large valve at the bottom of the cylinder is placed on seat a little without the cylinder, connecting freely with it, and the valve is large in order to allow the expelled air to escape instantly and without essential resistance and noise. This valve is made airtight by means of a leather or otherwise") [2]. It was to be opened in the BDC position of the piston and be closed at TDC ("opened when the piston arrives at the bottom of the cylinder and to remain open until the piston has risen to the top") [1]; ("in order that the cylinder may refill with air each time, the air valve must be opened when the piston arrives down, and must remain open till it has arrived up, and the explosion takes place") [2], this being immediately after complete combustion of the mixture in the cylinder ("the valve at bottom closes at the same time") [1]; ("a large valve at or near the bottom of the cylinder, which is then instantly allowed to close") [2].

It was operated by means of a linkage connected to the piston rod and held open by a dog, until the stream of gas flowing out of the valve after ignition of the mixture actuated the valve by help of a suitable mechanism. According to [1], this was accomplished by a small vaned wheel, according to [2] by a plate on which the gas flow impinged, and thus allowed the valve to return to its closing position ("a small rod that passes down by the side of the cylinder turns under the bottom and is connected to the lower end of the valve stem, so that when the piston rod descends, it bears on this, pushes down or opens the valve, a small dog at the bottom of the valve stem falls into a notch in the same, and suspends it in that position ... until the explosion takes place, the strong current of air driven out by the explosion strikes the vanes of a small wheel, disengages the dog from the stem of the valve, which then by a spring instantly rises to its place") [1];

("I plan a lever or spring, whose support is about midway between the center of the cylinder and the center of the valve, and when the piston reaches the end of its stroke it will press upon the lever, and start up and raise the valve high enough to be detained by a catch, which works into a notch on its stem ... that the valve may shut the moment the explosion takes place, the catch which suspends the valve must let go. ... the catch be connected with an arm, and supported at the angle between them on a pin so that the arm shall extend down to the current of air issuing from the valve, and putting a broad piece of metal or wood upon it to be acted upon by the blast, it will be driven off and unlatch the valve at the instant required") [2].

Fig. 7 shows the operating principle of the 1824 engine. The exhaust and air inlet valve EV=AV was opened at piston BDC (position I), and also was the mixture valve MV at a certain distance before piston TDC (position II) ("when the piston has risen 1/2, 3/4, or 7/8 of the length of the cylinder, the communication with the preparing box is to be opened") [1]. For this purpose, a stopcock was operated by the crank mechanism, as was also the ignition valve IV, but which was closed in a different manner ("I generally use a large stopcock for opening and shutting the communication between the box and cylinder which is worked by its connection with the pitman, which connection also opens the valve on the tube to inflame the mixture, but it is shut with a weight or spring") [1]. At the moment the valves EV=AV and MV closed at TDC (position III) ("the communication with the box is then closed") [1], the ignition valve IV was opened ("the ... prepared air within the cylinder and ... above the common air which occupies the lower part of the cylinder is inflamed") [1] and instantly closed again. The inflamed and expanding mixture should then clear the cylinder of air (position IV before the valve EV=AV was closed (position V) ("it explodes and by its great expansion clears the cylinder entirely of the common air; the valve at bottom closes at the same time") [1]. The cooling of the combustion gases should then produce as complete a vacuum as possible, which would permit the atmospheric pressure acting on the piston P to be utilized for mechanical purposes ("having the cylinder filled only with a highly inflamed vapor ... a condensation instantly follows, leaving a very perfect vacuum in the cylinder; the pressure of the atmosphere on the piston gives the power for use ") [1].

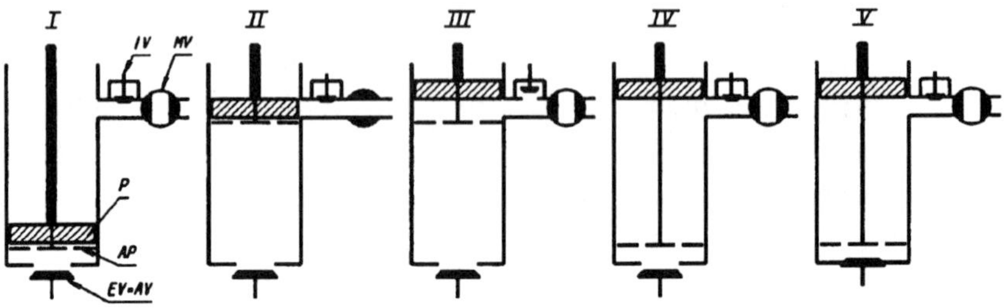

Fig. 7: Operating principle of the 1824 engine

An important component of all Morey engines was the "auxiliary piston" [2], also named "false piston" [1] or "charging piston" [3]. It was "made of metal" and arranged below the main piston ("its place is below the real piston") [2] and "should be fully perforated with holes" of about 3 mm diameter ("about 1/8 of an inch") [3]. According to [2], the rod of the auxiliary piston was guided in the rod of the main piston ("the rod of which will have a hollow to receive the rod of the other") and was held in it by means of a light spring ("whose stem is detained in the hollow of the main rod by a spring with just force enough to sustain its weight"). According to [3], it was guided by a stuffing and held by its friction ("it is fitted with a small rod, which moves through a stuffing box in the main piston, so tight that the friction will support its weight"). There is no indication, however, where this stuffing box was located, possibly in the rod of the main piston. The auxiliary piston AP was to be attached to the main piston P at BDC (position I) and should stop just below the mixture inlet, while the main piston moved on to its TDC ("I cause it to stop, when within from 1/4 to 1/8 of the height of the stroke of the working one") [1], which was likely to have been achieved by an appropriate stop. This was, however, not specifically mentioned until in [3] ("this charging piston is prevented rising higher than the vapor pipe, by its end striking against a stationary point in its ascent"). The space thus formed between the two pistons (position III) was to serve for measuring the mixture volume ("to measure the space to be occupied by the vapor") [2]. The expansion of the inflamed mixture drove the auxiliary piston down to its bottommost stop (position IV) ("the explosive mixture ... when inflamed, drives the light piston to the bottom") [1];

("on the explosions taking place it acts above or on the upper side of the auxiliary piston and drives it to the bottom of the cylinder, and the air before it") [2], and after the atmospheric pressure had likewise pushed the main piston down (position I), the auxiliary piston was reengaged to the main piston ("the pressure of the atmosphere brings down the working piston to the lower one, and is again attached to it as before") [1]; ("the piston then descends, and the rod or stem of the auxiliary reenters the bore of the piston rod to its spring which detains and lifts it and carries it up to its former place as the piston ascends") [2].

Fig. 8 shows schematically the pressure/volume diagram of the atmospheric working cycle as performed in Morey's engine. Air was drawn into the cylinder through the exhaust and air inlet valve EV=AV from point A at piston BDC and a cylinder pressure p_o corresponding to atmospheric as far as point B. Subsequently, fuel-air mixture was drawn in from point B and closing of the exhaust and air inlet valve EV=AV and opening of the mixture valve MV as far as the point C at TDC. After closing of the mixture valve MV and ignition of the mixture by opening and instantaneously closing of the ignition valve IV, the cylinder pressure dropped from point C as the piston moved down to point D at BDC. At this moment, the exhaust and air

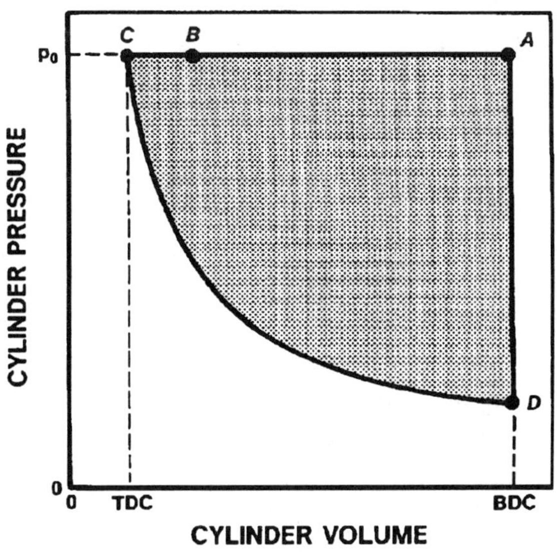

Fig. 8: Schematic atmospheric pressure/volume diagram

inlet valve EV=AV was opened, whereupon the pressure in the cylinder increased to the atmospheric pressure p_0 and to point A, and the engine was ready for the next working cycle.

The smallness of the area A-B-C-D representing the work performed in the cylinder illustrates the drawback of the atmospheric cycle, namely the very low attainable specific power output which is limited to a brake mean effective pressure (BMEP) less than atmospheric. That the figures achieved in actual engines were even considerably lower will be discussed in a later chapter. As Fig. 9 shows, the 1825 engine differed from that of 1824 solely in the design of the mixture valve MV, for which Morey now used "a poppet valve, through which the vapor passes at the proper moment into a horizontal pipe leading into the cylinder a little below the piston, when nearly up" [2]. That he retained this type of valve, as is discussed below, is a further indication that the undated manuscript [2] originated at a later date than the text [1], namely about 1825.

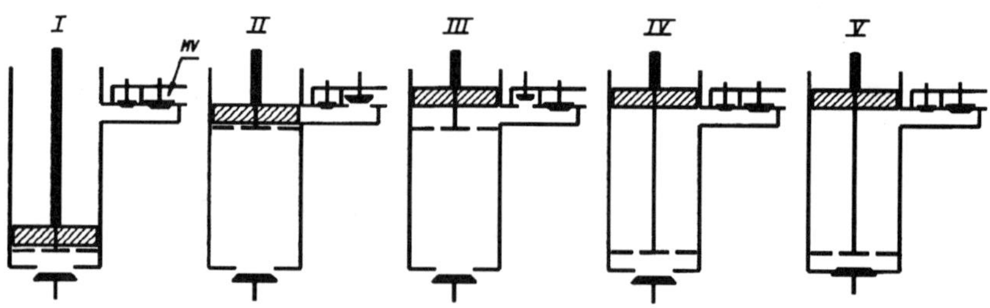

Fig. 9: Operating principle of the 1825 engine

For this state of development, Morey has supplied a description of the operating principle of the ignition valve. It was positioned in the mixture inlet pipe between the mixture valve MV and the cylinder ("In order to fire ... the charge of vapor in the cylinder, there is an opening outward on the pipe leading thereto between it and the valve before mentioned") [2], and it was "closed with a poppet valve" and opened at the moment the mixture valve closed ("opened at the instant the other is shut") in order for the mixture flowing out to get in touch with an ignition source ("whereupon a portion of the vapor issues and comes in contact with the flame of a lamp, or any other means of ignition") [2], and then it shut again immediately. The ignition and the mixture valve were operated in a suitable sequence by a mechanism connected to the crankshaft ("the obstructing valve and the igniting valve in the pipe leading from the evaporator to the cylinder may be moved in succession by any usual mechanical movements ... but that which is most convenient at present is one taken from the bar which connects the beam with the crank of the flywheel. From it a rod proceeds and accompanies its motion so as to move a small shaft which has starts and catches connected with it to lift and hold in succession these two valves") [2]; a method totally different from the one used in later years.

Fig. 10 represents the operating principle of the 1826 engine. What was novel was solely the application of separate valves for the air inlet AV and the exhaust outlet EV. The exhaust valve opened by itself immediately after ignition of the charge (position IV) and closed again instantly. It should have at least half the bore diameter and open outwards ("let the lower end of the cylinder have a valve, of at least one half its diameter, opening outwards") [3]. It was a hose of "thin soft leather of the same diameter of the working cylinder, and may be tied or fastened to the lower end of the cylinder ... so as to form a continuation of the same" [3]. Its lower end was "flattened, so as to bring its inner sides together for about four or five inches of its length, and kept in that position by light springs, attached to the two edges of the flat part, placing it in a position much like that of a bow and bowstring" [3].

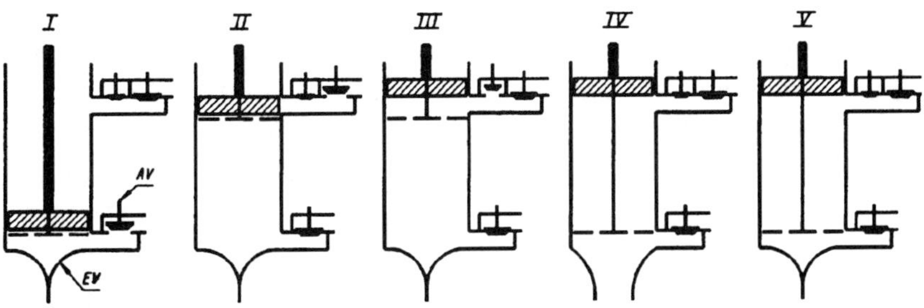

Fig. 10: Operating principle of the 1826 engine

In order to keep the atmosphere from forcing the leather valve into the cylinder, a plate supported at the bottom end of the cylinder and perforated with the maximum possible number of holes of 3 to 6 mm diameter was provided as an abutment ("this valve is supported or prevented from being driven into the cylinder by a plate of metal, of sufficient thickness to bear the pressure of the atmosphere, arched or raised outwards and perforated with as many holes of about 1/8 to 1/4 of an inch diameter as can well be made in it. The end of the cylinder ... forms the abutments to this arched plate") [3]. Morey retained this valve design, as is revealed in the eyewitness report [13] ("when the explosion took place in the cylinder a flame would shoot through a clap valve at the bottom of the cylinder").

The air valve AV, opening outwards and arranged on the side at the bottom end of the cylinder ("an air valve also, opening outwards is fixed to the side of the cylinder just below the piston when down"), was actuated by cams on the crankshaft, as were the ignition and the mixture valve ("they and the air valve, are worked by cams on the crankshaft") [5]. In contrast to the earlier used exhaust and air inlet valve EV=AV, the air valve AV was to be closed as soon as it arrived at the mixture inlet and the moment the mixture valve opened (position II) ("the air valve, which supplies the cylinder with air until the piston reaches the pipe from the preparing box, it then closes") [3].

PARTICULAR DESIGN FEATURES

The mixture was carried "from the evaporating box ... to the cylinder ... by means of a bellows" [2]. For the engine start, the bellows was to be operated manually ("the engine to be set in motion will require that the bellows be worked by hand a few moments") [2] in order to deliver an inflammable mixture which was to be ignited in the cylinder to first warm up the engine ("the explosive air may be blown into the cylinder, and inflamed, to warm it, and commence the operation") [6]. Subsequently, the bellows was to be operated by the engine itself, although it is not known how this should be done. Neither Hazard's patent drawing [5], nor any of Morey's manuscripts reveals any such mechanism. It can be assumed, however, that a cam on the crankshaft was used for this purpose. This air supply system appears to have been very important to him for reasons that it caused the engine to come to a standstill as soon as the bellows stopped working, and not least, that the fuel consumption instantaneously ceased in this case ("and the moment the bellows stops, the engine stops, and all evaporation of the liquors ceases, and of course all expense") [1].

Morey initially assumed the fuel-air mixtures to be "lighter than common air" [2] and that, therefore, no mixing took place in the cylinder ("it mix not with the common air, which being the heaviest it will not do, but remain floating upon it") [2]. This assumption holds true only for the air portion of the—warm—mixture, but not for the vapors of alcohol and turpentine with their respective 1.6 and 4.7 times higher densities compared to air of the same temperature. In addition, the auxiliary piston arranged below the main piston was to act as a kind of partition to prevent a dilution of the fuel-air mixture and thus avoid any adverse effect on the engine output ("a means of preventing the charge of vapor from mingling with the common air in the cylinder") [2]; ("a partition between the compound air in the cylinder and the common atmospheric air, thereby preventing the mixture of so great a proportion of atmospheric air in the cylinder as to lessen the effect of the explosion") [3]. It was, in fact, extremely important for the effectiveness of the engine that the air in the cylinder, of between 7/8 and 11/12 of the swept volume, was kept separate and completely expelled ("keeping the explosive mixture separate from the atmospheric air in the cylinder and driving out the latter without making use of it in the combustion") [5].

Whether, however, the auxiliary piston was guided and sealed in the rod of the main piston, as is described in [2], or whether it was designed as shown in the detail reproduced in Fig. 11 from Hazard's patent drawing [5], is a question which must remain unanswered. Morey had to learn that the use of the auxiliary piston was not beneficial on all operating conditions. Although it permitted using lean mixtures and burning less fuel with the same power output ("when the temperature is low smaller charges will produce the same effect, if a thin metallic plate of the same diameter as the piston, be introduced into the cylinder") [3], it was advisable to stop the auxiliary piston when rich mixtures were applied, even though this involved some loss

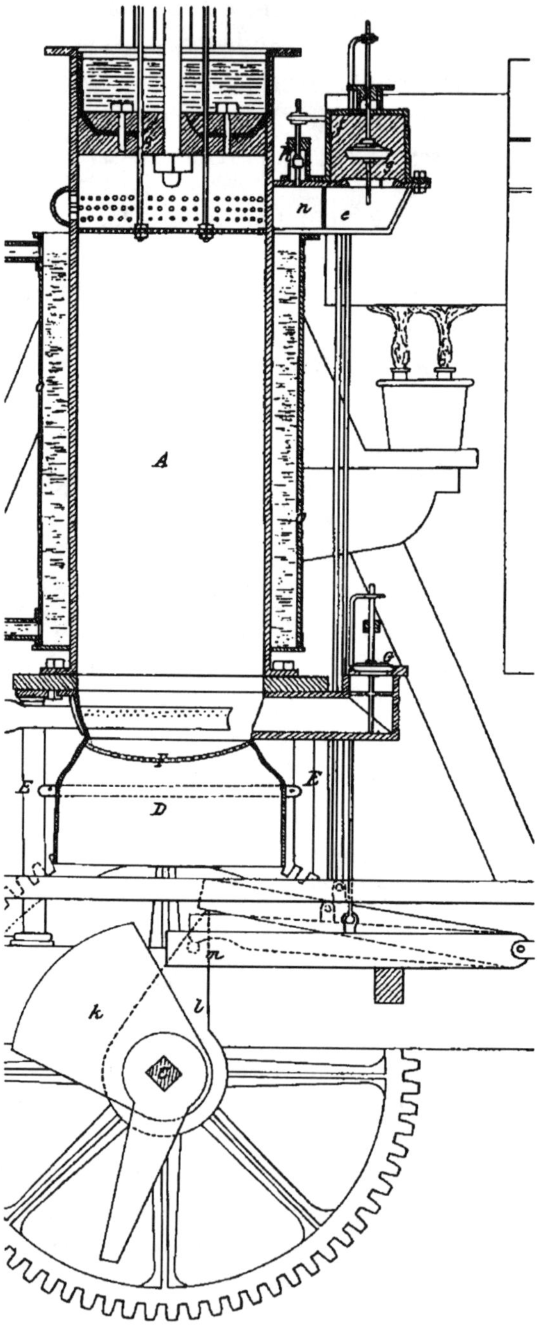

Fig. 11: Cross section of the 1826 engine [5]

in fuel efficiency ("when the temperature of the box is so high as to give off too great a proportion of vapor, the engine works better, by stopping entirely the working of the charging piston, but not with so good economy") [6].

During the development work of the year 1825 described in [2], Morey introduced a cooling of the exhaust and air inlet valve by means of "water allowed to enter the cylinder to keep the lower valve cool." This is again a further indication that the undated text [2] originated later than the manuscript [1] in which no cooling is mentioned. With this cooling method, "a small tube is inserted in the opposite side of the passage the outer end of which will be in a vessel of water, so that the instant the vacuum is formed, a jet of cold water will strike the underside of the valve.... This water runs out by another pipe having a valve opening outward, while the cylinder is refilling with air" [2]. The design of the internal cooling of 1826 by means of injecting "a small stream of cold water near the bottom, which also keeps the arched plate and leather valve cool" [3], is not clear. According to the British patent, the water injection was controlled by a "cock ... worked by a cam on the crankshaft" [5], and it may be assumed that this was Hazard's idea, since Morey never made mention of this.

Completely new in the 1826 engine was the water jacket of the cylinder (Fig. 11). Curiously, this aspect is not mentioned in the U.S. patent [3], but is in the account [6] written only a few months later ("keeping the lower part of the cylinder cool, by surrounding it with water"), and also in the British patent [5] ("surrounding it with water in the jacket"). The idea was to have the cooling water circulated, as can be derived from the arrangement of a water inlet and outlet.

The operational safety of the engine was a particular concern of Morey's. Beyond the wire gauze at the evaporator outlet, he provided for an additional wire mesh in the mixture supply line as a further measure to prevent a flashback in the event of untimely closing of the mixture valve ("the precaution may be taken to put a wire gauze, across the pipe leading from the evaporator, notwithstanding the stop valve cuts off the communication") [2]. Also, he recommended a further wire mesh to be installed in the mixture inlet pipe between the mixture and the ignition valve ("wire gauze should also be placed between the vapor and inflaming valve, between the vapor and inflaming valve, to prevent explosion in the box, should the vapor valve not close in time") [3]. As regards unintentional overfilling with fuel-air mixture, Morey had found that this operating incident was harmless ("as to the cylinder, should that from any cause become entirely filled with the explosive mixture, and exploded, as it has a valve at the bottom, standing open or but barely balanced by a light spring, of nearly or about 1/2 of the diameter of the cylinder, it is unquestionably impossible to burst it or force out the piston") [1], based on the consideration that no damage to the cylinder was likely because of the only very low flow resistance which the exhaust valve offered to the combustion gases and also on the basis of various tests ("a cylinder ... I have repeatedly filled entirely with the exploding mixture, and inflamed it, without the least appearance of injury, and to make it sure that the whole cylinder was filled with the prepared air, it was filled near the top, and is inflamed at the valve on the bottom") [1]. In addition, he saw in the downward opening valve an effective means of preventing injury to people ("when the valve opens downwards, I believe I have guarded effectually against a possibility of any persons being ever injured by any explosion that can possibly take place") [1].

A particularly noteworthy design feature, first mentioned in the U.S. patent [3], was the actuation of the valves by means of "cams on the crankshaft." That [6], in a similarly worded sentence talks of "beams" instead of "cams" is likely the result of a reading or printing error. As can be seen from Fig. 11, this mechanism involved cams, rocker arms, pushrods, and poppet valves, in a combination which is still common today in similar form.

It is improbable that Morey ever used electrical ignition, as Charles Edgar Duryea (1861-1938)—together with his brother J. Frank (1870-1967), one of the pioneers of automobile development in the U.S.—has claimed (54). Although already in [1] mention is made of the advantages of this igntion method ("this power is produced [if the vapor is inflamed by the electric spark] without fire at all, when the temperature of the weather ... is only about 80°"), and practically identical statements are found in [3] and [6], surprisingly though not in [5], no source gives the slightest indication of an actual use. Rather, the eyewitness report [13] reveals that Morey retained his method with ignition valve and a flame placed in front of it ("a small hole opposite to which a lamp was burning").

HIS IDEAS OF ENGINE COMBUSTION

Morey was aware of the fact that hydrogen-air mixtures take up a much greater space than those composed of alcohol, turpentine, and air ("it is known that it takes about three times as much hydrogen gas to form an atmosphere that will inflame, as it does to prepare one that will explode") [1]. Also, he proceeded on the basis that the combustion of both types of mixture produced largely the same effect ("the effect of the explosion of the prepared air from the liquors appears to be much the same as that of hydrogen gas and atmospheric air") [1]. From this, he concluded that alcohol-turpentine mixtures were the most advantageous for producing as strong a vacuum as possible.

He also knew that the attainable vacuum is dependent only on the extent to which the fuel-air mixture increases its volume during combustion ("producing by explosion a ... more extensive vacuum than the space the prepared air previously occupied") [3]. Apparently, though, he was not aware that the increase in volume at the practically constant pressure which can be assumed to have existed in his engine depends solely on the ratio of the temperatures of the burnt and the unburnt mixture. This is revealed by his assumption that a mixture filling of 1/12 of the cylinder was adequate to achieve its appropriate evacuation ("it ought to be less than 1/12") [1], which would have necessitated enlarging the gas volume by a factor of 12, and thus a temperature of about 3400°C, which was not feasible in the engine. The mixture filling of 1/10 to 1/8 stated in [2] ("it will be found sufficient to occupy one tenth or one eighth of the cylinder with the vapor"), which requires combustion temperatures of about 2800 and 2200°C, respectively, is much closer to reality.

Morey's ideas regarding the combustion of alcohol-turpentine mixtures, in particular the significance of the water added to them, were extremely curious. To his opinion, "steam from the decomposition and recomposition of the water" [1] affected the engine process, and he assumed that during fuel combustion, water was dissociated ("the water ... is decomposed yielding its oxygen and hydrogen") [2], which afterwards would recombine ("forming anew into water") [2]. The view was presumably based on erroneously interpreted experience which he had gained from his experiments with his "American water burner" patented in 1817, but also on knowledge acquired from the literature.

That textbook information played a considerable role in his thinking is revealed by the use of terms clearly obtained from such volumes, as for instance that of the "boiling point" [2]. The same can be derived from statements which obviously reproduce unfamiliar knowledge. A characteristic example for the latter case is his use of the term 'caloric' ("steam ... when it loses its caloric ... is condensed into water") [2] which today would be translated with heat content, but in those days meant the 'caloricum.' This had been conceived by Laurent Lavoisier (1743-1794) and was regarded—through to the 19th century—as a massless substance being responsible for heat and combustion phenomena (similar to the 'phlogiston' which, originally conceived by Johann Joachim Becher (1635-1682), was disclosed in 1710 by Georg Ernst Stahl (c.1 660-1734) and proven to not exist by Lavoisier in 1772) (55). That Morey also concerned himself with chemical matters is illustrated by his description of the hydrogen production by means of iron and sulfuric acid ("hydrogen gas is formed by decomposing water by a process hastened by the use of iron and sulphuric acid, the former having a strong affinity for the oxygen, sets the hydrogen free") [2]. This example reveals that he based his remarks on comparatively recent literature, for the term 'oxygen' was by no means common knowledge in the twenties of the 19th century (even if the word can be reliably traced back to 1786, 12 years after Joseph Priestley [1733-1804] and Karl Wilhelm Scheele [1742-1786]—independent of each other—had discovered the corresponding chemical element).

DESIGN AND PROCESS VARIANTS

Beyond his standard engine concept, Morey considered some other designs and operating methods, as are described in [1] to [3] and [6], but not in [5].

In two instances, this involved the reinvention of devices which had been described by Jean de Hautefeuille as early as 1687 (56). The first was a method in which by means of the combustion of fuel—in Hautefeuille's case, this was gunpowder (41)—a gas pressure was to be produced over a column of water which thus was to be forced to a corresponding altitude ("by the explosion, the air is compressed over a column of water, to force that to a great height or distance") [1]; ("when the object of the engine or power is to raise water only, the explosion may be made in one vessel to drive the air into another partly occupied with water, which is driven out and upward by a tube to any height equal only to the pressure or elasticity of the air thus produced") [2]. The second reinvention was the reversed process in which, by burning fuel and cooling the combustion gases, a vacuum was to be produced in a vessel, by means of which water could be raised, even if to only a very limited height, namely to less than 9.8 m ("reversing this operation, if a vacuum be formed at an elevation of 32 feet in a vessel connected by a pipe with water below, it will immediately rise and fill the vacant vessel") [2].

Another idea, which Morey was unlikely to have realized, was to produce hot combustion gases, by means of which air was to be compressed which should pass through a tube to some kind of expansion device ("I cause the explosion to compress in a reservoir a quantity of atmospheric air, and make use of that compressed air for working an engine similar to those of a ... high-pressure steam engine") [1]. He thought the principal benefit being that the expansion machine could be operated at some distance from the gas producer and at locations where there was a fire risk ("this mode makes it perfectly safe ... on account of fire as the compressed air may be led in tubes any distance from where there is any fire, before it works the engine") [1].

At a very early date, Morey gave consideration to overcoming the greatly restricted power output of his engine, resulting from the atmospheric working principle, by making it at least partially direct-acting and thus more powerful. His setup is schematically represented in Fig. 12 and is characterized by an upward opening valve PV in the main piston ("I make a valve in the piston to open upwards") [1] and a cylinder closed at the top, not expressly mentioned in [1]. The auxil-

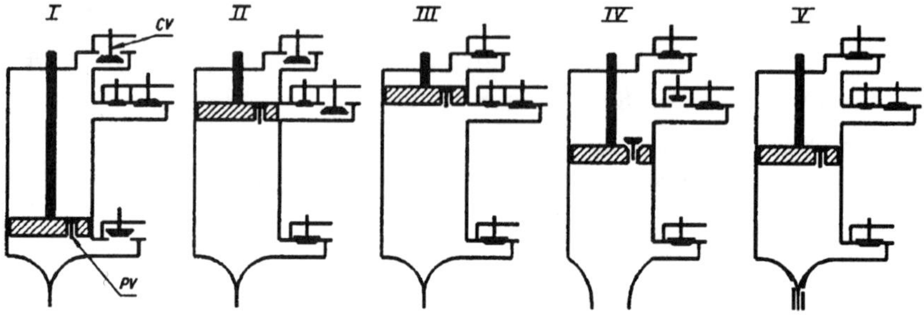

Fig. 12: Operating principle with partial direct action

iary piston has been omitted from the figure, in order to provide a clearer illustration and also because Morey did not make mention of its use for this variant. With the piston moving upward between positions II and III, the cylinder was filled with fuel-air mixture which was conveyed into the space above the piston as this moved down, until the mixture was ignited once the piston was halfway down (position IV) ("I ... fill the cylinder below the piston with the prepared air, and when the piston is at the height of its working stroke, it is only about halfway up the cylinder; the explosion then takes place"). At this point, the piston valve closed, and the piston was forced down by the pressure built up in the space above ("the effect is, that the air above the piston is nearly doubled in quantity, its elasticity or force is also greatly increased, by a great increase of its temperature, it now reacts on the piston with much force, while a vacuum below adds greatly to the effect"). Due to the transfer of most of the mixture into the space above the piston, it is not possible for any appreciable vacuum to have occurred below the piston, which is why the exhaust valve must have been opened slightly between positions IV and V, as is alluded in Fig. 12 by the broken line, to permit expelling of the air from the bottom part of the cylinder.

A similar description is found in [2], with the express reference to the achievable increase in cylinder power ("there is... another modification of the engine which ... increases the power of one cylinder") and to the closed cylinder above the piston ("the cylinder be closed airtight at top ... and ... be so prolonged upward as to afford an air chamber above the piston when up"). The desired effect could not have been achieved with the only vaguely described setup, though, as Morey has omitted to mention the downward movement of the piston to the point of ignition at half the stroke. In contrast, however, he did mention for the first and, interestingly, last time the need to have an inward opening valve CV for emptying the space above the piston, which implies renewing the air therein contained ("this method requires valves to let in air again below, and let out the compressed air ... above") [2]. A prerequisite for effective employment of direct action, though, was the use of inward opening ignition and mixture valves, not mentioned at any point. In the U.S. patent [3], the same operating principle is described, this time once again incomplete in that little mention is made of the cylinder closed at the top or of the valve required for emptying the space above the piston ("I have made a valve in the piston to open upwards, and fill the cylinder below the piston with the prepared air, and when the piston is about halfway up the cylinder, it is at the height of its working stroke, the explosion then takes place, the effect is that the quantity of air above the cylinder is nearly doubled, its elasticity or force is also greatly increased by a great increase of its temperature, it now reacts on the piston while the vacuum below adds much to the effect").

Fig. 13 shows a schematic representation of the pressure/volume diagram of Morey's engine operating principle with direct action. Air is aspirated through the valve AV from point A at piston BDC and a cylinder pressure p_0 equal to atmospheric, and with the combustion chamber valve CV open, until point B. From point B and closing of the air valve AV and opening of the mixture valve MV, fuel-air mixture passed into the space below the piston until point C at TDC. During the subsequent downward movement of the piston with the combustion chamber valve CV and the mixture valve MV closed, this mixture was transferred into the space above the piston as a result of the piston valve PV opening automatically, until point D. Ignition occured at this moment by briefly opening and immediately closing the ignition valve IV, as a result of which the pressure above the piston increased to the point E. Then the piston valve PV closed automatically, and the pressure decreased during the downward movement of the piston to the point F at BDC. The combustion

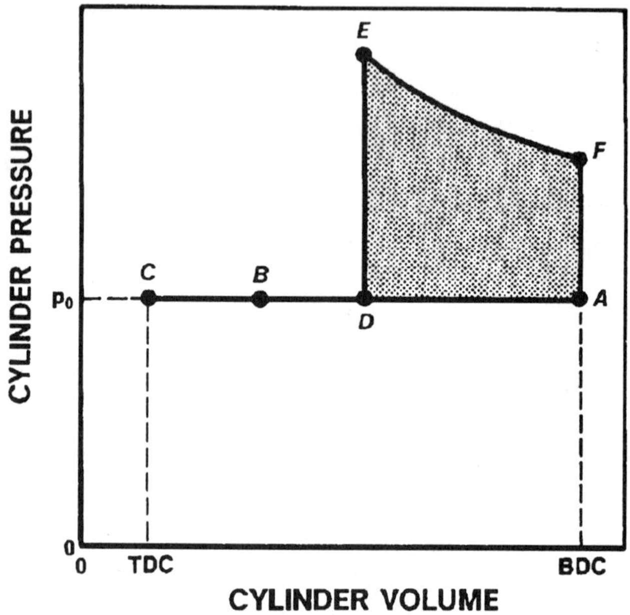

Fig. 13: Schematic pressure/volume diagram with direct action

chamber valve CV was opened at this moment, whereupon the pressure in the cylinder dropped to atmospheric pressure p_0 and to the point A, and the air valve AV was opened again for the next cycle. It is evident that the mixture valve MV and the ignition valve IV must open inward or be greatly preloaded to permit a pressure higher than atmospheric in the combustion chamber above the piston, which Morey is likely to have known, but did not mention. The pressure trace shown in Fig. 13—which corresponds to that in the compression-less engine which Joseph Etienne Lenoir built more than 30 years later—clearly reveals the drawback of this operating principle, namely that a significant part of the stroke is used for filling the combustion chamber with mixture, which is to the debit of the enclosed diagram area D-E-F-A and thus at the expense of the attainable BMEP and which rules out a large expansion ratio required for a high engine efficiency.

A further experimental variant of the atmospheric engine is described in [3] and [6]. Tubes of about 10 mm diameter but unknown length were attached to the main piston, each of which was to enter 19 mm wide tubes of a condenser attached to the bottom end of the cylinder ("I have attached to the working piston, thin tubes open at top, about 3/8 of an inch diameter, each one of which is directly over and enters one of the tubes of the condenser, attached to the bottom of the cylinder, which tubes were about 3/4 of an inch diameter"). The mixture filled "the upper part of these tubes and the spaces around them" and burned less violently ("these tubes prevent all violence in the explosion"), an effect which is likely to have been due to the increased heat dissipation caused by the enlarged combustion chamber surface. This idea also does not seem to have been pursued.

A further experiment described in [3] and [6] indicates that Morey was not satisfied with the vacuum produced in his engine. In this case, he had attached to the bottom end of the cylinder, up to 40 tubes of about 19 mm diameter and 20 cm length, which were surrounded by cold water and were to thus permit the production of a more complete vacuum ("I have sometimes attached 30 or 40 tubes about 3/4 of an inch diameter to the bottom of the cylinder, about 8 inches, and kept them surrounded with cold water, the more instantly and perfectly to perfect the vacuum") [3]. As Morey did not make any more statements regarding the success of this measure, it can be concluded that he abandoned this idea, probably due to insufficient effectivity.

THE POWER POTENTIAL OF THE ENGINE

It can be assumed that Morey did not perform any measurements of the power output of his engines. Subsequent calculation is impossible, not least because their actual cylinder dimensions and the number of strokes performed per unit of time are not known. Nevertheless, an estimation of power and fuel consumption figures based on measured data from other atmospheric engines is tried.

Stating that "the power of the vapor engine may be calculated on the same principles as that of the steam engine" [2], Morey provided the following calculation rule: "Multiply the area of the piston in inches, by the weight of the atmosphere 14 3/4 pounds, then by the length of the stroke and the number of strokes per minute and divide by 32,000, the number of pounds which a horse is supposed capable of raising one foot in a minute." This rule is misleading, as the comparison with the calculation formula given below shows. In this, P is the engine output in horsepower (HP), A the piston area, S the piston stroke, and SPM the number of strokes per minute.

$$P\ [HP] = (14\ 3/4\ [lb/in^2] \times A\ [in^2] \times S\ [ft] \times SPM)/33{,}000$$

On the one hand, the stroke S must be inserted in ft in order to obtain for the power P the result in the unit ft.lb/min, corresponding to modern horsepower definition (550 ft.lb/s = 76.04 kgm/s = 1.341 kW). On the other hand, the conversion coefficient resulting from the conversion of seconds into minutes, being 60 times the defined horsepower, should read 33,000 lb/min.

His calculation example, according to which "an engine whose 2 cylinders should be 1 foot diameter and 3 foot stroke would be a 16 horsepower" [2], shows that he knew how to apply the rule correctly, as he must have inserted the stroke S in ft in order to obtain the power stated. He obviously assumed the engine speed to be 50 SPM, if one proceeds from a conversion coefficient of 32,000 (a factor of 33,000 would result in about 52 SPM). This engine speed harmonizes with the lower limit of the range of the number of mixture charges which, according to his statement in [3], an evaporator could supply. Such a high power output would have been unattainable, though, as the rule stated does take into account neither mechanical nor thermodynamical losses. Morey is likely to have been well aware of the former, yet he was not able to estimate the mechanical efficiency of his engine. However, it was not only he who was unaware of the impossibility of performing thermodynamic cycles without any losses. Although this principle had been established by Nicolas Léonard Sadi Carnot (1796-1832) in 1824, it had not yet become common knowledge, as his book was published initially only in a small private edition (57).

An estimation of the power output and fuel efficiency attainable with Morey's engine can best be derived from a comparison with those of the atmospheric engines which Nikolaus August Otto (1832-1891) and Eugen Langen (1833-1895) developed in 1867 (58) and subsequent years. These engines can be regarded as representing the highest attainable technical state of this engine type. This conclusion is underscored by the fact that around 5000 examples—an unusually large number for conditions in those days—were manufactured industrially with power outputs from 1/2 to 3 HP. Measurements conducted on such an engine with a cylinder diameter of 150 mm and a stroke of 980 mm operating at 42 SPM, corresponding to a mean piston speed of 1.37 m/s, resulted in an output of 0.64 HP corresponding to a BMEP of 0.40 bar, and an efficiency of 13.7%. An output of 0.35 HP and an efficieny of 15.4% were determined at 20 SPM corresponding to a mean piston speed of 0.65 m/s (19). Investigations on a considerably larger engine with a cylinder diameter of 317 mm and a stroke of 1030 mm, revealed an output of 2 HP, again corresponding to a BMEP of 0.40 bar, and an efficiency of 11.2% at 28 SPM, corresponding to a mean piston speed of 0.96 m/s (18). With a swept volume of 81.3 dm^3, the last-mentioned engine is likely to have been close to Morey's in size. With the aforementioned calculation example from [2] (2 cylinders with 1 foot bore and 3 feet stroke), this would have had a total displacement of 133.4 dm^3, with which it would have been possible to achieve a maximum power output of only 4 HP with a BMEP of 0.40 bar corresponding to the two aforementioned engines, instead of the 16 HP calculated by Morey. Based on the cylinder bore of about 9 or 10 inches stated in the eyewitness reports [12] and [13] and a stroke/bore ratio of 2.68 obtained from Hazard's patent drawing [5], the engine would have had a total displacement of around 60 dm3 and less than 2 HP.

Morey came considerably closer to the truth with his estimation of the fuel consumption of his engine, which was likely based on practical experience, in which he assumed that it could be operated with 3.79 dm^3 each of turpentine and drinking alcohol for hours ("how long one gallon of spirits of turpentine, with one gallon of whiskey or other cheap spirit will last, is presumable from the smaller quantities heretofore used, and may be estimated at hours, though this must be referred to more experience") [2]. With a specific combustion enthalpy of the turpentine of 41.8 kJ/g (38.2 kJ/cm^3) and of the drinking alcohol, assumed to contain 40% by volume of ethanol, of 9.2 kJ/g (8.4 kJ/cm^3), the energy content calculated for the two gallons (7.57 dm^3) of fuel mixture is 190 MJ (25.1 kJ/cm^3). Assuming an efficiency of the engine of 13%, which has been verified as being attainable in

(18) and (19), this means a specific fuel consumption of 0.22 gal/HP-hr (1.1 dm^3/kWh), which, with the power output of between 2 and 4 HP to be assumed, would have permitted the engine to operate with the two gallons of fuel mixture for 2 or 4 hours, respectively, quite in agreement with Morey's estimate.

ENGINE APPLICATIONS

In his efforts to market his invention, Morey demonstrated the suitability of his engine for various applications, in some cases as part of more or less public demonstrations.

The first might have been the one in New York, of which Thomas Ewbank (1792-1870) (7), author of a widely distributed book on the raising of water (30), reported: "In 1826, Mr. Morey ... patented an explosive engine, and soon after exhibited a large working model in ... New York" [12]. His indication that the demonstration took place "soon after" the issue of a patent, might point to the year 1826, as was assumed in (5). It should be noted, however, that Ewbank's book was not published until 1856, almost thirty years after the event described, a time scale in which even several years might well be considered to be "soon after." It is also conceivable that the demonstration did not take place until 1828 and that the engine displayed in New York was identical to that built in Philadelphia in the winter of 1827-1828 [7]. Apparently, Morey conducted several demonstrations, as is revealed by the fact that Ewbank "took several opportunities to examine" it. It is not known, however, who was the organizer or the intermediary of these events and who was intended to be referred to. That nothing is said to indicate that the engine provided power for any kind of mechanical purpose might mean that it was only operated under idling conditions and that this demonstration is not identical to the "application to stationary purposes" mentioned in [8] and performed in the winter of 1828-1829. This assumption is underscored by the statement that a heated "small tin dish" [12] was adequate for preparing the fuel mixture, whereas Morey's evaporator would have been needed for full load operation.

A further demonstration may have taken place at the beginning of the year 1828 in Philadelphia, as is revealed by an account by Thomas P. Jones (53): "Mr. Morey is also here, with the design of completing one of his explosive engines" [7]. In this case it is also not known whom he wished to interest in his engine. Whether the "application to stationary purposes" [8], which, according to the date indicated, took place in the winter of 1828-1829, was likewise the subject of a demonstration, is not known.

According to a letter of Orford's Minister Sylvester Dana [9], Morey set off—in the company of his nephew Edward Bissell (1800-1857)—on October 15, 1829 for Baltimore "to make one more vigorous effort, to sell his patent right." There, he hoped to be given the opportunity of demonstrating the suitability of his engine for locomotive purposes ("I can but hope and trust the ensuing winter will see the engine well applied to a carriage on a railroad") [8], which Morey regarded as the last task for him to solve ("when that is done, I should think I have done my part"). These endeavors—aimed at the Baltimore & Ohio Railroad Company, which had just offered a prize for the best locomotive (59)—were obviously in vain, and led him to head for Philadelphia. There, too, the installation of his engine in a locomotive of one of the three local railroad companies—the Philadelphia & Columbia, the West Philadelphia, and the Philadelphia Germantown & Norristown—might have been his aim. That the engine was demonstrated on other occasions was probably due to the "Messrs. Rush and Muhlenberg" and to "Mr. Garrett" mentioned in [8].

James Rush and David Muhlenberg, the latter a grandson of Heinrich Melchior Mühlenberg (1711 -1787), "Patriarch of the Lutheran Church in America" (60), were sons-in-law of Oliver Evans (1755-1819) (61), who was known for his work on the use of steam engines in boats and the development of the high-pressure steam engine (32). As owners of a foundry and mechanical engineering works in Philadelphia and manufacturers of steam engines (32), Rush and Muhlenberg may have considered acting as the producers of the Morey engine, if there was sufficient customers' interest. Philip Garrett, a jeweler and watchmaker, was a prominent member of the Philadelphia Fire Engine Company, one of the many rival fire brigades split up into "fire engine and hose companies" (32). From him, Morey may have expected connections to manufacturers of fire fighting pumps and thus the possibility of motorizing them, a hope quite justified, as is shown below.

The exact date of the demonstration is unknown. The course of events was reported by George Escol Sellers (1808-1899), author of momentous memoirs of the history of engineering (32). Although he described the events of those days, at which time he was just 21 years old, at a very much later date, namely after 1883, his statements are regarded as being very reliable. That he mentioned only in passing an application of Morey's engine for "propelling a bateau on the Schuylkill" [13]—a 211 km long river flowing through Pennsylvania in a south-easterly direction and converging at Philadelphia with the Delaware River—leads to believe that he was not present himself at that event. His eyewitnessing seems to have begun when Morey's engine "was ... brought to the fire engine shops." This refers to a company founded about 1818 by Sellers' father Coleman and Abraham L. Pennock (Sellers & Pennock, 231 Market Street) for the production of the "Hydraulion" [13], "a combination fire engine and hose carriage in one" (32). Based on Sellers' knowledge, Morey had no "plans to attach" the engine "to the fire engine pump," but "he only proposed to furnish the driving power," in order "to demonstrate the practicability of running the engine to a fire without the aid of men with the long ropes," who at that time hauled the fire engine to the scene of the fire. This demonstration was not conducted with a fire engine, however, but with a light, four-wheeled vehicle with a springy platform, which Morey built himself ("he mounted a kind of buckboard carriage"), for which, as Sellers presumed, he used "a set of light small wheels" of a "Perkins garden engine." This was "an oval upright tub on two wheels, with wheelbarrow handles and legs, with a small single-chamber pump in it" (32) to fetch water from public cistern pumps for the garden. The manufacture of this had been the cause of a partnership of Sellers' father with Jacob Perkins (62) (Perkins & Sellers), which ended in 1818, and which had proven an economic failure. Below the engine, Morey built "a pan containing water" to keep the vehicle platform from burning. He probably had brought from Orford the parts needed to transmit the engine power to the driving wheels. According to local tradition, Morey is said to have already conducted running tests in his workshop, which are said to have ended in an accident ("the engine ... got out of control during a test and nearly or quite smashed its way out through the side of the shop") (1). That he thereupon did in no way "shelve his plans" for a motor vehicle, as is claimed in (4), is illustrated by the driving test described in Sellers' words:

"After Mr. Morey had made what he thought satisfactory yard tests, he had the gate onto Market Street opened that he might mount the seat he had provided, guide the machine and run out Market Street which was then a common road not having been paved with cobble stones although the side gutters were paved with brick, but the narrow paved sidewalks were not curbed.

In starting the machine, somehow Mr. Morey when about mounting his seat lost his hold, tripped and fell flat. As he gathered himself up calling "Stop her, stop her," which no one seemed ready or knew how to do, the thing ran across the street, through the gutter, over the sidewalk and turned a somersault into the brick yard where the clay had been cut away several feet lower than the sidewalk, a complete wreck.

This was the end of the vacuum engine" [13].

As Fig. 14 shows, this short and unfortunate, although nevertheless memorable, run began in the yard of 231 Market Street at the corner of Sixth Street and ended in the clay pit of a brick yard on the opposite side, which today are both parts of the Independence Mall. It was the first witnessed—even if involuntarily unmanned—automobile run in America, and indeed the second at all. The first one, extending over a distance only slightly longer (about 50 m), had been made by Isaac de Rivaz in 1813 in Vevey on the Swiss side of the Lake Geneva, likewise with an atmospheric engine (49).

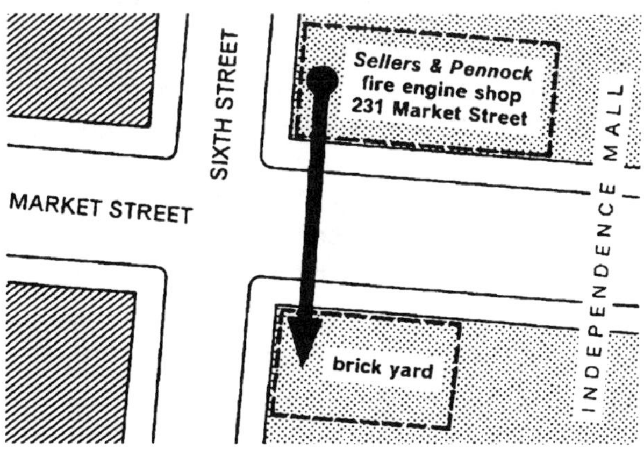

Fig. 14: Morey's run in Philadelphia

Although the fate of Morey's engine was thus settled in a very abrupt manner, this by no means "was the end of the vacuum engine" [13]. Sellers, who did not make this statement until the 80s of the last century, had obviously not heard of the atmospheric engines developed by Otto and Langen and manufactured by the Gasmotoren-Fabrik Deutz AG in Cologne (63). This fact is quite surprising since these were sold in the U.S.—by Schleicher, Schumm & Co., Philadelphia—and one of them had been exhibited there in 1876 at the Centennial International Exposition (5).

From [8] it is known that, prior to his travel to Baltimore in October 1829, Morey conducted tests for using his engine for powering a boat ("the application of the 'new power' to a boat"). It had a length of about 5.8 m ("about nineteen feet") and a width of approximately 1.7 m ("5 1/2 feet"), and the engine was to take up only about 46 cm ("about eighteen inches") of the stern. This length means that the engine must have been mounted transversely to the boat's longitudinal axis, an arrangement particularly suitable for driving paddle wheels on both sides of the boat. Morey's figures regarding its speed of 11.3 to 12.9 km/h ("between 7 and 8 miles per hour")—or even 13 to 14.8 km/h if nautical miles were meant—cannot be based on experience. The same holds true for the estimate that a suitably modified engine would be capable of propelling a boat of twice the capacity at a speed of at least 16 or 18.5 km/h, respectively ("the same engine may unquestionably, when in better hands, be made to drive one, properly constructed, of twice the capacity at least ten"). In the same way, Morey's optimistic statements regarding the running properties and fuel consumption of the engine ("It will now run as regular as any that are driven with steam, and with very little expense") were based only on expectations.

It should be noted that Morey's boat trips were not the first using an internal combustion engine. He had two forerunners, namely the brothers Joseph Claude and Joseph Nicéphore Niepce, the latter one of the fathers of photography (24), who had conducted runs with a boat on the Saône at Chalon in France as early as 1807, using an engine of remarkable design, named "Pyréolophore" (5). Similarly, Samuel Brown had demonstrated a boat to the British Admirality on the Thames near London in January 1827 (64).

Morey is said to have built in the early 20s (54) "a large boat" of about 6 m ("full twenty feet") "in length, painted white, with a red streak and black gunwhale" (4), named "Aunt Sally" and powered by a steam engine. In 1874, Dr. Willard Hosford (1805-1882), Morey's family physician, mentioned a boat of Morey's which "was sunk about 50 years" earlier (4). Based on this remark, the sinking was dated to "about the year 1825" (5) or to "1821" (54), combined with the presumption that the boat involved was the "Aunt Sally." According to local tradition, it was sunk in Fairlee Pond and many years later had been seen lying on the bottom of what today is Lake Morey (3).

These statements are not conclusive since Morey had conducted all his boat trials on the Connecticut River, as is confirmed by eyewitness reports (65), until he moved to Fairlee in 1832. They do make sense, however, if the "Aunt Sally" and the boat mentioned in [8] are considered identical, which suggests itself on the basis of the similarity of the hull length of "fully twenty" (4) and "about nineteen feet" [8], respectively, and if it is further assumed that the sinking occured in the 30s, in other words 40 years before Dr. Hosford's statement. In this case, the boat lying at the bottom of Lake Morey might be the "Aunt Sally" converted in the fall of 1829 from steam engine to internal combustion engine power [8], as is assumed in (3). This supposition meets with local tradition according to which Morey used his boat for pleasure trips on Fairlee Pond long after his move (4).

The prospect of recovering Morey's motorboat caused the New Hampshire Antiquarian Society to search the lake in 1874, unfortunately without any success, and attempts made in 1982 and 1984 using sonar technology did not provide more satisfactory result. Nevertheless, a planned undertaking using magnetometric methods is linked with the hope of laying hands on one of Morey's engines and thereby finally clarifying the question of its design and thermodynamic properties.

MOREY'S INVENTIVE PRIORITY

On the basis of present-day general basic concepts of justice, Morey was without doubt entitled to legal protection for his engine. Yet, it is of interest to examine what was in fact novel on his invention. To consider Morey's invention as a prior art of that of all modern reciprocating engines, as Duryea did—"all ... make the power stroke ... down. Thus correctly did Morey sense the internal combustion engine" (54)—would mean bestowing on its inventor an honor which he would not have claimed for himself.

That Morey was not the first to design an engine in the U.S., can be derived from two remarks of Thomas P. Jones. In 1833, when discussing a patent application relating to an engine powered with explosive substances [10], he remarked that he had "assisted in the performance of a number of experiments ... upwards of twenty years ago," i.e., prior to 1813, on an engine, probably to be powered with gunpowder, but he did not note whether it was atmospheric or direct-acting and who was the engine builder. The location of the experiments also cannot be deduced as Jones' place of residence at that time is not known; he still may well have been living in New Bern, N.C., his first place of residence after he had immigrated into the U.S. "as a young man" (7), i.e., at a date unknown (33). The second remark relating to another precursor of Morey's engine is found in [7]. According to this, "in the year 1814, a model of an engine ... was placed in" Jones' "hands ... with a view to his instituting some experiments with it," but he had to interrupt these tests because of his "removal ... to Virginia," Again, the engine builder is unknown, as is the location of the tests which Jones conducted. From the context, it is understood that it was an atmospheric engine, which is why Morey cannot be regarded as the creator of the first American engine of this type.

Neither the operating principle nor the automatic control of the valves and the ignition permitting a continuous operation of the engine, both of which had already been used in 1820 by William Cecil (52), were in fact novel ideas, but rather the combination of certain design features, which individually were indeed state of the art. Thus, cams, push rods and rocker arms had been common since medieval times (66), poppet valves had long been known, leather hoses were used as exhaust valves by Christian Huygens in his gunpowder engine 200 years before (41), and electrical ignition was used successfully by Isaac de Rivaz on his engine a decade earlier. In these circumstances, it is not surprising that Hazard's patent [5] claimed as an inventive novelty with respect to the engine components only "the peculiar construction of the flexible valve" and the auxiliary piston for "keeping the explosive mixture separate from the atmospheric air in the cylinder." Interestingly, the cylinder cooling by means of a water jacket, which was actually a novel design feature, was not part of the claims.

That the U.S. patent [5] claimed the use of low-boiling fuels as novel did not correspond to the state of the art to the extent that Robert Street had stated in 1794 that materials such as "spirits of tar or turpentine" (42) were suitable engine fuels. Novel in fact was the use of fuels vaporized outside the engine, for the mention of "vapor of alcohol or any ardent spirits, or of oil of turpentine and the essential oils" (38) by Samuel Brown from 1826 were not prejudicial to novelty. Whether the discovery of the inflammability of fuel vapor-air mixtures was really novel for contemporary science, as Morey assumed, is of no consequence here.

Rather, what is of importance is that his heated evaporator was the first device used on an internal combustion engine for mixing fuel vapor and air (5,54). This is why Morey has to be considered the inventor of the heated surface carburetor, an invention which so far has been ascribed to Luigi de Christoforis for the year 1841. This most important invention of Morey's was not superseded until about 60 years later by Wilhelm Maybach's (1846-1929) nozzle carburetor (5). To an equal extent, Morey was the first to use liquid fuels for automotive purposes, for George Medhurst's (1759-1827) engine from 1800 (67,68) was designed to use gunpowder (69), and Isaac de Rivaz had used gas for his run in 1813.

Conclusion

It is not known when Morey abandoned his engine work. No further news exists regarding this subject after the winter of 1829-1830. Thus, it may be that the demonstration described in [13] meant the end not only of his efforts to sell his patent rights but also of all development activities. This assumption is supported by his statement passed down in [9] that he wished to let the matter rest in the event of a failure "and return to Orford, to spend his days in quiet." This occured no later than his move to Fairlee (1832) as is revealed by the fact that he commenced development work at that date on a novel lamp, as is confirmed by a letter to Prof. Silliman of 1833 (70). He must have been disappointed, especially after his engine initially, in 1826, had been extraordinarily optimistically judged—probably by Thomas P. Jones ("should no unforeseen difficulties present themselves ... it will be the greatest improvement which has been made for many years") [4]. Particularly tragic was that his prediction regarding future traffic—"there will be little use of horses for that purpose" [1]—and the consequent change in "the commercial and personal intercourse of the country" was to prove much later to be truly prophetic. Unfortunately, however, this did not hold true for the use of his own engine.

What Morey had quite correctly realized was the particular suitability of internal combustion engines for vehicle propulsion ("this principle is positively well adapted to the purpose of locomotive engine") [2], a view shared also by the editor of the Franklin Journal and American Mechanics' Magazine ("particularly in its application to locomotive engines") [4]. In reading these statements, it should not be forgotten that the term 'locomotive' as used in those days referred not only to railroad propulsion but, in the original sense of the word, equally to any engine-powered transport device.

It is no longer possible to determine for what reasons Morey's efforts to market his engine failed, but it was a mistake on his part to turn to the wrong kinds of customers. Although the specific fuel consumption of his engine was better than that of a comparable steam engine, a fact of which Morey does not seem to have been aware, it was not an attractive proposition for boat and locomotive builders because of its small power output. It was in this field that the steam engine maintained its place in competition with the internal combustion engine far into the present century.

What he did not realize was the fact that the atmospheric engine was suitable only for meeting certain gaps in demand which the steam engine was unable to meet or where its use, involving as it did, considerable weight, space, and expense in terms of time, costs and manpower, could not be justified. Thus, the later, outstanding commercial success of the atmospheric engines of Otto and Langen (58)—which because of their great height and noise were nicknamed "rattling monsters"—can only be explained by the fact that they marketed them as meeting the power demands of small stationary power users. It was not until the 80s of the last century that direct-acting, high-speed engines were developed—by Karl Benz (1844-1929), Gottlieb Daimler (1834-1900) and Wilhelm Maybach—which first supplied the specific power outputs required for automotive use (5).

A disadvantage for Morey was that his invention did not really gain proper recognition on the part of science. Thomas P. Jones, who in this field also spoke for the Franklin Institute, doubted "whether ... an engine operated upon by an explosive mixture of gas, or vapor, will ever be constructed, which will successfully compete with the steam engine" [7]. In this, he was not quite wrong, whereas his assessment of the prospects of direct-acting engines proved to be a misjudgment: "The objections to these, appear to us insuperable, for ... the practical difficulties... are likely to insure, the abandonment of all such attempts."

Morey's engine was obviously too much ahead of its time to be properly assessed, with the consequence that it rapidly fell into oblivion. In Europe, its inventor's name has been mentioned merely once [11] for which reason it did not affect later developments, as

Duryea has alleged: "No one can tell the far reaching influence of this wonderful man" (54). In fact, Morey's engine, a model of so-called "Yankee ingenuity" (5), has remained a solitaire which is nevertheless worthy of its place in the history of engineering, and until a monument is unveiled at some date in honor of its inventor and the first American automobilist, may the present paper serve a modest substitute.

Acknowledgements

Particular thanks are owed to Mrs. Alice Doan Hodgson, Orford, N.H.—"who has been researching Morey most of her adult life and has defended him since school days when any teacher said that Fulton invented the steamboat" (59)—who provided the author with a wealth of details regarding Morey and his home village and was untiring in answering even the oddest questions. The same is true also of Patrick M. Raher, Washington, D.C., who, with great patience and care, supplied most varied information and helped with a detective's nose in procuring documents. Without this valuable support, the present paper would never have been accomplished.

References

1. *Samuel Morey Memorial Bridge Dedicated.* The United Opinion. Bradford, Vt., July 1,1938

2. Getman, F.H.: *Samuel Morey, a Pioneer of Science in America.* Osiris, 1 (January 1936) pp. 17-36

3. Hodgson, A.D.: *Orford, New Hampshire - A Most Beautiful Village.* Orford, N.H., 1978

4. Hodgson, A.D.: *Samuel Morey - Inventor Extraordinary.* Orford, N.H., 1961

5. Cummins, C.L.: *Internal Fire.* Lake Oswego, Oreg., 1976 (Warrendale, Pa., 1989)

6. Morey, S.: *On Heat and Light.* American Journal of Science and Arts. 2 (May 1820) 1, pp. 118-132

7. *Dictionary of American Biography.* 20 vols. New York - London 1928-1936

8. Jones, T.P.: *For the application to machinery of the Power resulting from Explosive Compounds - Henry Rodgers, New York, June 29.* Journal of the Franklin Institute. 7 (1833) 6, pp. 409-410

9. Jones, T.P.: *Explosive Gas-Engine - Henry Rodgers.* Mechanics' Magazine. 21 (1834) pp. 110-111

10. Verity, C.W.; Quigg, D.J.: *The Story of the U.S. Patent and Trademark Office.* Washington, D.C., 1988

11. Schubarth, B.: *Repertorium der Technischen Literatur - Die Jahre 1823 bis einschließlich 1853 umfassend.* Leipzig 1856, p. 547

12. *Methoden zur Bereitung einer Explosionsmischung, und Anwendung derselben als Triebkraft bei Maschinen, worauf Erskine Hazard ... in Folge einer ihm von einem im Ausland wohnenden Fremden gemachten Mitteilung und eigener Zusätze sich am 12. August 1826 ein Patent erteilen ließ.* Dinglers polytechnisches Journal. 8 (1827) 11, pp. 377-381

13. *Recent Patents. To Erskine Hazard ...; in consequence of a communication made to him by a certain foreigner residing abroad, and additions made by himself, for an Invention of a Method or Methods of preparing Explosive Mixtures, and employing them as a moving power for Machinery.* The London Journal of Arts and Sciences. vol. 13 (1827) 77, pp. 1-6

14. Richard, G.: *Les nouveaux moteurs à gaz et à pétrole*. vol. 2, Paris 1892

15. Sass, F.: *Geschichte des deutschen Verbrennungsmotorenbaues von 1860 bis 1918*. Berlin - Göttingen - Heidelberg 1962

16. Tresca, M.: *Sur l'invention et l'avenir des machines à gaz combustibles*. Annales du Conservatoire Impérial des Arts et des Métiers. 2 (1861/1862) pp. 121-152

17. Donkin, B.: *A Text-Book on Gas, Oil, and Air Engines; or Internal Combustion Motors without Boiler*. London 1894

18. Clerk, D.: *The Gas, Petrol, and Oil Engine - Thermodynamics of the Gas, Petrol, and Oil Engine, together with Historical Sketch*. vol. 1, London 1910

19. Güldner H.: *Das Entwerfen und Berechnen der Verbrennungskraftmaschinen und Kraftgas-Anlagen*. 3. ed. Berlin 1920

20. Farrell, G.: *Captain Samuel Morey - Who Built a Steamboat Fourteen Years before Fulton*. Manchester, N.H., 1915, p. 11

21. Morey, S.: Will, March 21, 1843. Probate Office Fairlee, Vt.

22. Allen, H. B.: *The Franklin Institute of the State of Pennsylvania*. Transactions of the American Philosophical Society. (1953) vol. 43, pp. 275-279

23. Morey, S.: *An Account of a New Explosive Engine, Generating a Power that May be Substituted for that of the Steam Engine.* American Journal of Science and Arts. 11 (October 1826) pp. 104-110

24. *Dictionaire de Biographie Française.* 13 vols. Paris 1933-1975

25. *Dictionary of National Biography.* 63 vols. London 1885-1900 (22 vols. 1959-1960)

26. Hulls, J.: *Machine for Carrying Vessels or Ships Out of, or Into any Harbour, Port, or River; Against Wind and Tide, or in a Calm.* British Patent of December 21, 1736

27. Hulls, J.: *Machine for Carrying Vessels or Ships in a Calm.* London 1737

28. Morey, S.: *An Account of a New Explosive Engine, Generating a Power that May be Substituted for that of the Steam Engine.* The Franklin Journal and American Mechanics' Magazine. 2 (August 1826) pp. 115-119

29. *Morey's New Vapour Engine.* The Franklin Journal and American Mechanics' Magazine. 1 (April 1826) p. 252

30. Ewbank, T.: *Descriptive and Historical Account of Hydraulic and Other Machines for Raising Water, Ancient and Modern: With Oberservations on Various Subjects Connected with the Mechanic Arts: Including the Progressive Development of the Steam Engine.* New York 1856

31. Sellers, G.E.: Memoirs Book IV. pp. 63-64 (Peale-Sellers Collection, Library of the American Philosophical Society, Philadelphia, Pa.); in (32)

32. Ferguson, E.S.: *Early Engine Reminiscenses (1815-1840) of George Escol Sellers.* Smithsonian Institution, Washington, D.C., 1965, Bulletin 238

33. Fowler, F.: *Memoir of Dr. Thomas P. Jones.* Journal of the Franklin Institute. vol. 103 (1890)1, pp. 1-7

34. *Papers of William Thornton.* vol. 5, Library of Congress, Washington, D.C., 1917

35. Thornton, W.: *Short Account of the Origin of Steam Boats, written in 1810, and now committed to the press.* Washington, D.C., 1814

36. Davenport, N.: *The United Kingdom Patent System - A Brief History with Bibliography.* London 1979

37. *Gas Engine Carriage.* Mechanics' Magazine. (June 1826) pp. 79-80

38. Brown, S.: *Certain Improvements on my former Patent, dated December 1, 1823, for an Engine or Instrument for Effecting a Vacuum, and thus producing Power, by which Water may be Raised and Machinery put in Motion.* British Patent A.D. 1826 ... No. 5350

39. Brown, S.: British Patent A.D. 1823 ... No. 4874

40. Brown, S.: U.S. Patent of March 2, 1824

41. Hardenberg, H.: *An Historical Overview of Gunpowder Engine Development -1508-1868.* American Society of Mechanical Engineers. Internal Combustion Engine Technical Conference, Dearborn, Mich., 1989

42. Street, R.: *Method to Produce an Inflammable Vapor Force by means of Liquid Air, Fire, and Flame, for communicating Motion to Engines, Pumps, and Machinery* British Patent A.D. 1794 ... No. 1983

43. Lebon d'Humbersin, P.: Addition of August 25, 1801 to French Patent No. 356 of September 28, 1799

44. Niepce, J.C.; Niepce, J.N.: French Patent No. 409 of April 3, 1807

45. Niepce, J.C.; Niepce, J.N.: *Brevet d'invention de dix ans pour une machine appelée pyréolophore, dont le principe moteur est l'air dilaté par le feu.* Description des machines et procédés spécifiés dans lies brevets d'invention, de perfectionnement et d'importation dont la durée est expireé. vol. 7 (1824) pp. 47-63

46. Niepce, J.C.: *Certain Improvements in the Means of Propelling Boats and other Vessels, and which Improvements are also applicable to Machinery of various Descriptions.* British Patent A.D. 1817 ... No. 4179

47. Cayley, G.: Notebook. pp. 95-100; in (48. pp. 42-44)

48. *Aeronautical and Miscellaneous Note-Book (ca. 1799-1826) of Sir George Cayley* Cambridge 1933

49. Michelet, H.: *L'inventeur Isaac de Rivaz.* Martigny (Switzerland) 1965

50. de Rivaz, I.: French Patent No. 731 of January 30, 1807

51. de Rivaz, I.: *Machine à combustion.* Descriptions des machines et procédés spécifiés des brevets d'invention de perfectionnement et d'importation dont la durée est expirée. vol. 9 (1826) pp. 182-184

52. Cecil, W.: *On the Application of Hydrogen Gas to produce a moving Power in Machinery; with a Description of an Engine which is moved by the Pressure of the Atmosphere upon a Vacuum caused by Explosions of Hydrogen Gas and Atmospheric Air.* Transactions of the Cambridge Philosophical Society. vol. 1. pt. 2 (1820) 14, pp. 217-239

53. Jones, T.P.: *Some Remarks upon Explosion and Vapor Engines.* The Franklin Journal and American Mechanics' Magazine. 5 (January 1828) pp. 18-19

54. Goodwin, K. R.; Duryea, C.E.: *Samuel Morey, Precursor of Motor Power Development.* The Vermonter. 36 (May 1931) 5, pp. 135-142

55. Segré, E.: *From Falling Bodies to Radio Waves - Classical Physicists and Their Discoveries.* New York 1984

56. de Hautefeuille, J.: *Pendule perpetuelle, avec un nouveau balancier; et la maniere d'élever l'Eau par le moyen de la Poudre à Canon, et autres nouvelles inventions, contenuës dans une Lettre de Monsieur de Hautefeuille, écrite à un de ses Amis.* Paris 1678

57. Carnot, L.S.: *Reflexions sur la puissance motrice du feu et sur les machines propres à developper cette puissance.* Paris 1824 (Paris 1978)

58. Otto, N.A.; Langen, E.: *Atmosphärische Gas-Kraftmaschine.* Dinglers Polytechnisches Journal. 168 (1867) 10, pp. 90-97

59. Hodgson, A.D.: Letter to the author of August 21, 1990

60. Frick, W. K.: *Heinrich Melchior Mühlenberg - Patriarch of the Lutheran Church in America.* Philadelphia, Pa., 1902

61. Bathe, G.; Bathe, D.: Oliver Evans. Historical Sociery of Pennsylvania, Philadelphia, Pa., 1935

62. Bathe, G.; Bathe, D.: *Jacob Perkins - His Inventions, His Times, and His Contemporaries.* Historical Society of Pennsylvania, Philadelphia, Pa., 1943

63. Goldbeck, G.: *Kraft für die Welt - Hundert Jahre Klöckner-Humboldt-Deutz AG.* Essen 1970

64. *Mr. Brown's Gas Vacuum Engine.* Mechanics' Magazine. (February 1827) pp. 82-84

65. Davis, J.L.: *Capt. Morey and His Boats.* The United Opinion. Bradford, Vt., August 14,1931

66. Klemm, F.: *A History of Western Technology.* Cambridge 1964

67. Medhurst, G.: *A New Improved Method of Driving Carriages of all Kinds, without the Use of Horses, by means of an Improved Æolian Engine, and which Engine may also be applied to various other Useful Purposes.* British Patent A.D. 1800 - No. 2431

68. Hardenberg, H.: *George Medhurst's Schießpulverkraftwagen.* Automobiltechnische Zeitschrift 93 (1991) 6, pp. 378-379

69. Hardenberg, H.: *"Die Kraft des Schießpulvers zum Wohl und Nutzen der Menschen zu wenden..." - Materialien zur Geschichte des Schießpulvermotors.* (in the press)

70. Morey, S.: Letter to Professor Silliman. (Pennsylvania Historical Society) 1933

Appendix

1. Morey's "Papers prepared for Publication Octr 1824, but not sent"

Manuscript preserved in the Baker Library, Dartmouth college, Hanover, N.H.

Sir. - Accident has given to the mechanical world a new power, some account of it follows. Perhaps it may be worth noticing in your journal. I have found that if the vapor or steam of water, or water and alcohol and spirit tuurpentine is mixed with a great proportion of atmospheric air that the air is rendered highly explosive, that the explosion is followed by a very perfect vacuum. I have also found, that if a current of atmospheric air is blown rapidly over the surface of those liquors, at a temperature only to about blood heat, and but a short distance, that the air is rendered very explosive. To prepare and perfect the preparation in great quantities for use in a small compass aspossible, we take a box, made of common tin plates, say about one foot diameter, and seven or more inches deep, with horizontal divisions once in two or three inches, the space between those divisions is divided by vertical partitions, into spaces half an inch or more apart, in a zigzag or spiral form. When the temperature of the box and liquors is about blood heat, atmospheric air is to be blown into the lower space in the box on one side, it passes back and forth in contact with the liquors, to the opposite side, when it passes up the lower horizontal division, into the next space above it, then again takes its zigzag course to the opposite side, and will pass, if more divisions, through the whole in the same manner, into the space between the upper division and cover, being about one inch deep, fit for use. To use it, and to produce the power, we take a cylinder and piston, similar to those used for steam engines, to the bottom of which we attach a valve to open downwards, of about from 1/4 to 1/2 the diameter of the bore of the cylinder, this valve is to be opened when the piston arrives at the bottom of the cylinder and to remain open until the piston has risen to the top, but when the piston has risen 1/2, 3/4 or 7/8 of the length of the cylinder, the communication with the preparing box is to be opened, so that the air, driven through the preparing box, and which in its course through the same has become highly explosive, passes into the cylinder below the piston, to fill the same as fast as the piston moves up, until the piston has risen to the height of its working stroke, the communication with the box is then closed, the exploding or prepared air within the cylinder and next the piston, and above the common air which occupies the lower part of the cylinder, is inflamed, it explodes and by its great expansion clears the cylinder entirely of the common air, the valve at bottom closes at the same time, having the cylinder filled only with a highly inflamed vapor much of which is evidently red hot steam from the decomposition and recomposition of the water, a condensation instantly follows, leaving a very perfect vacuum in the cylinder, the pressure of the atmosphere on the piston gives the power for use. To inflame the explosive mixture within the cylinder without loss, I cause the explosive air to pass through a tube having three or four zigzag turns, 6 or 8 inches long each, after it is inflamed, before it reaches the cylinder and explodes, then by having time to close the valve where it is inflamed, saving the whole effect of the explosion below the cylinder, forming a larger and more perfect vacuum for use. I generally use a large stopcock for opening and shutting the communication between the box and cylinder which is worked by its connection with the pitman, which connection also opens the valve on the tube to inflame the mixture, but it is shut with a weight or spring. The following method is found very convenient for working the large valve at the bottom of the cylinder by a small rod that passes down by the side of the cylinder, turns under the bottom and is connected to the lower end of the valve stem, so that when the piston rod descends, it bears on this, pushes down or opens the valve, a small dog at the bottom of the valve stem falls into a notch in the same, and suspends it in that position, while the piston is rising and until the explosion takes place, the strong current of air driven out by the explosion strikes the vanes of a small wheel, disengages the dog from the stem of the valve, which then by a spring instantly rises to its place. This particular mode of producing the power promises, I should think, much both from its great simplicity and from being able (in a sense) to produce an unlimited power with little or no violence, as the explosion meets with little or no resistance, by the largeness of the valve at bottom, except that of the surrounding atmosphere. Say for instance only 1/8 of the cylinder is required to be filled with the explosive mixture (and from experiments it ought to be less than 1/12) to drive out or clear the cylinder of air by its expansion, it will in effect be equal to a force of eight atmospheres. The following are some of the many methods I have made use of for producing the power. Sometimes I cause the explosion to compress in a reservoir a quantity of atmospheric air and make use of that compressed air for working an engine similar to those of a double-stroke high-pressure steam engine. This mode makes it perfectly safe and convenient on account of fire as the compressed air may be led in tubes any distance from where there is any fire, before it works the engine. The power from the vacuum in this case may be made to force in other quantities of air, or applied to other useful purposes. Sometimes by the explosion, the air is compressed over a column of water, to force that to a great height or distance.

Sometimes I make a valve in the piston to open upwards and fill the cylinder below the piston with the prepared air, and when the piston is at the height of its working stroke, it is only about halfway up the cylinder, the explosion then takes place, the effect is that the air above the piston is nearly doubled in quantity, its elasticity or force is also greatly increased by a great increase of its temperature, it now reacts on the piston with much force, while a vacuum below adds greatly to the effect.

Sometimes I have a light or false piston attached to the working one, and cause it to stop, when within from 1/4 to 1/8 of the height ofthe stroke of the working one, and fill the space between the two with the explosive mixture, which when inflamed drives the light piston to the bottom, clears the cylinder of air, a valve at the bottom closes, a vacuum is formed, the pressure of the atmosphere brings down the working piston to the lower one, and is again attached to it as before. This mode, too, has the advantage of working in the large way with great economy and but little force from the explosion. The condensation, when necessary, may be effected as that of steam is. As this engine has no boiler nor anything of any weight as a substitute for it, as it can be made very light and compact, as it will give any desired power, will require next to nothing in point of weight or dense to continue the power, when that of four or five horses for hours, and as it requires no fire except a lamp, I see no reason why it may not (in addition to the uses to which steam is applied) be applied to the greatest advantage in driving carriages on good roads and railway, and particularly for giving what seems to be much wanted direction and velocity to balloons. Or the waste hydrogen gas of the balloon, or a reservoir of gas (which in some cases for such uses may be preferable to the vapor) may be used instead of the explosive vapor passing it with a sufficient quantity of atmospheric air through the preparing box, and applying it in the same manner to the machine for producing the power. What makes the discovery the more promising for these and many other purposes is that from many experiments there appears to be no doubt about that the cylinder may be made of sheet brass or copper and even of tin plates, that the piston can be hung with leather so as to move airtight and without friction, and without injury from the inflamed vapor. This power is produced (if the vapor is inflamed by the electric spark) without fire at all, when the temperature of the weather in the shade or in the sun is only about 80°. I have found that the air driven but once across the box only one good diameter in the zigzag direction, and with all the velocity and pressure of a common smith bellows is rendered highly explosive, a proof how easy, how cheap and how very readily this power is produced, but passing it twice or more it will work, not only at a lower temperature, but we can regulate the mixture as we please, by governing the quantity of air by stops, let into the lower or upper divisions of the box. When I say that a box of this size (which need not cost over two dollars, nor weigh more than six or seven pounds, and will last for years) will hold enough in the lower part to work an engine of a four or five hours, I set it lower than what I verily believe may be expected from it. With this engine, when the valve opens downwards, I believe I have guarded effectually against a possibility of any persons being ever injured by any explosion that can possibly take place. The preparing box is so constructed, from top to bottom, that it is little else than a continued but zigzag tube, which can, should the flame ever reach there, only explode one after another, (there being always a sensible chute, at each turn in the tube) but even of that there is little or no risque, as all the opening in the box can be covered by wire gauze of the fineness directed by Sir Humphrey Davy. The box may too have a safety valve of any size which need not be loaded at all if it weigh but two or three ounces. As to the cylinder should that from any cause become entirely filled with the explosive mixture, and exploded, as it has a valve at the bottom, standing open or but barely balanced by a light spring of nearly or about 1/2 of the diameter of the cylinder it is unquestionably impossible to burst it or force out the piston. Besides, this power is always limited in its effect and can be always easily met and contracted. A cylinder nearly one foot diameter, having eighteen inch stroke, made of tin plates, I have repeatedly filled entirely with the exploding mixture, and inflamed it, without the least appearance of injury, and to make it sure that the whole cylinder was filled with the prepared air, it was filled near the top, and is inflamed at the valve on the bottom.

It appears not to be essential, as to the proportion of each liquor put in, for the air invariably mixes with, combines, absorbs, or takes up nearly the same proportion of each, vary the proportions in the box as you may. That is, if water and spirit of turpentine are put in, in any proportion, the air takes up equal parts of each. If alcohol is added, it takes up nearly two parts of that, two of water, and one of the spirit of turpentine. If alcohol and spirit of turpentine only are put in, the air takes up nearly eight parts of the former to one of the latter. All these natural mixtures, if I may so say, when mixed with a very great proportion of atmospheric air, render it highly explosive, if their temperature is only about 80°. When raised considerably higher, the mixture in the cylinder will inflame, but without explosion making or forming however a very good vacuum. If raised still higher, it will not burn at all within the cylinder. At about 70° the spirit turpentine refuses to be taken up by the air or at least in suffcient quantities to give energy to the explosion. To retain its exploding power, (if the temperature is considerably increased above blood heat) it will require an increase in the quantity of air driven into a space in the box, above the lower one, or that containing the liquors. The vapor of alcohol, after passing the box, explodes without the spirit turpentine, but with less effect. This explosive air in passing through a half inch tube takes up sensible time, whereas the time in which a cubic foot will explode is insensible. We avail ourselves also of this fact to close the valve when the vapor is inflamed, before it reaches the cylinder, and is often found very convenient in experiments. When I use a cylinder without boring, I have found as yet no better way to make the piston move airtight, than to enclose it in a hoop of thin metal, that nearly fits the cylinder, from two or three to five or six inches deep; within that hoop, I put one within another, two or more cylinders of leather, made tight to the piston at bottom, and projecting some inches above the hoop. This cylinder of leather is pressed moderately against the sides of the cylinder by springs, or by some light elastic stuffing. From the fact now well known, that a flame cannot pass a very small space, it will readily be seen that it cannot possibly reach the leather between the hoop and sides of the cylinder, and as the cylinder is exposed every stroke both above and below the piston to the atmospheric air, there can be no apprehension of the leather, being ever injured from the heat of the cylinder or piston, which may too always work under oil or water. In this way the greater the pressure on the piston the surer it will be to move airtight. It is not necessary that the cylinder should have a temperature but a little above blood heat, if any.

Any person may satisfy himself of the fact, of the air being so readily prepared for producing a power, by proving a small tin plate box or even a few feet of inch tube in zigzag form. If the latter, drop in at one end a few drops of the liquors, raise the temperature as directed, blow with a common hand bellows through the tube, and it will be forced the air is rendered explosive. Or if a box is formed, blow in at the bottom or lower space, and receive the air in time, no matter what, if a proper tube, apply a flame, you will see how readily explodes. If 1/8 or 1/12 part of a cylinder 3 or 4 inches diameter is filled with the prepared air, and inflamed, it will expand through the whole cylinder. Now as it is known that it takes about three times as much hydrogen gas to form an atmosphere that will inflame, as it does to prepare one that will explode, it shows the great saving that must result in forming the exploding one, which is about two to one, but in addition to that, taking the advantage as we do, of causing the explosion to clear the cylinder of from 12 times or more of the space occupied by the explosive mirture, it gives an advantage over that of inflaming the vapor in the cylinder of not much odds of twenty to one. The effect of the explosion of the prepared air from the liquors appears to be much the same as that of hydrogen gas and atmospheric air. But I have found the former inclining to occupy a much greater space, and the more readily and spontaneously to form a greater and more perfect vacuum. This I have thought resulted from the act of the decomposition and recomposition of water. That much water is recomposed in the explosion, it appears very evident, by inflaming the prepared air in a glass tube. Water is deposited over the surface, so it is also in the cylinder. This water is likely every explosion converted instantaneously into steam, contributing to increase the act and perfecting the vacuum. The odds in the expense of procuring the two materials for many purposes and in many situations, must be, I should think, often fully and never less than ten to one, in favor of the prepared air. And this too can often be had conveniently in situations where the hydrogen gas cannot be provided to the air as prepared in a carriage or any other situation without the inconvenience of much heat. An engine of a ten horsepower can be put in motion in less than a minute, and the moment the bellows stops, the engine stops, and all evaporation of the liquors ceases, and of course all expense. Is there not some reason to expect that the discovery will greatly change the commercial and personal intercourse of the country. There is good reason I trust to conclude that transportation on good roads or railroad may be done much cheaper as well as quicker than by locks and canals, besides having the great advantage of being done, much of it, in the winter a time much the most convenient to the farmer. In their personal intercourse, if it should be generally thought most prudent to continue their intercourse on the earth's surface, yet I think there will be little use of horses for that purpose.*

[*There are two more pages which are not reproduced here, since they are not concerned with Morey's engine, but with sound phenomena experienced with flames burning in tubes]

2. Morey's Undated Specification Draft

Manuscript preserved in the Baker Library, Dartmouth College, Hanover, N.H.

To all persons to whom these presents shall come, be it known

that I, Samuel Morey, Orford in the County of Grafton and State of New Hampshire, a citizen of the United States, have discovered and made a new and useful improvement in mechanical power which I denominate the vapor engine, to distinguish it from the steam engine and the gas engine, and claim letters patent according to the laws of the United States. My said discovery and invention is described and specified in the words following viz.

Specification of Samuel Morey's invention denominated the vapor engine viz. I have by a long course of experiments discovered, found out and ascertained that the commingled vapors of water of alcohol and of spirits of turpentine, obtained at a low temperature viz, about 80° of Fahrenheit, will be an explosive vapor; when made to mix and chemically combine with air under cover of a vessel properly constructed for the purpose of mixing, preparing and blending the particles thereof. And that when the said vapor

is fired and exploded in a cylinder under a piston, it...*
air out thereof, or expels the portion of the at...
therein, and the air being prevented...
of a valve at the bottom of the cyl...
and instantaneously produced, run...
pressure of the atmosphe...
manner of the atmo...

And I have also found that modifications of the means will also produce the explosive vapor though less perfect. Alcohol and spirits of turpentine, having in them naturally a portion of water will also at the proper temperature produce it. But at a higher temperature, the vapor will rather burn than explode.

[*bottom right corner of the manuscript damaged]

And here it may be proper to remark that vapor differs from steam in that the former is formed at a lower temperature or below the boiling point, and is not at all or very moderately elastic, while on the contrary, steam has elasticity and force enough to expel and take place of the atmosphere, and has the property of being condensed on the application of cold. Vapor floats on and mingles with the air. Steam is lost and changed to vapor on coming into the open air when it looses its caloric and is condensed into water. Also that hydrogen gas is essentially dfferent from vapor and from steam. It is formed by decomposing water by a process hastened by the use of iron and sulphuric acid the former having a strong affinity for the oxygen, sets the hydrogen free. The hydrogen gas is inflammable. The principle, therefore, of the hydrogen gas engine is essentially different from the vapor engine

...s which I have discovered to be explosive at a
...ature easily produced, as herein subsequently
...form and blend but would naturally
.....nically too slowly for operation
... an excitement or impulse
... .apor of the spirits
... the object of the preparing
... current of air over
the surface of the mixt liquors, in such manner as to give it time to evaporate enough thereof and by so obstructing the air cause to become loaded, to mix and blend the particles in the most intimate manner possible.

Therefore I make a vessel of tin, or copper or other metal, or substance, into which I put the said spirits with water, which vessel need not have more area than three times that of the cylinder of the engine. And I have found the following method of contructing the interior thereof to be effectual, though susceptible of various modifications. Within the preparing vessel or evaporator I make several horizontal divisions, by means of partitions, and divide these chambers by vertical partitions into long compartments communicating at alternate ends; and these vertical partitions may have a range of small holes in them. If one or two chambers are thus partitioned into passages, one or two more horizontal partitions above them may be perforated with holes, in rows, but the rows of holes not opposite, but under the spaces between those above them, in order to commingle the vapors more intimately as they are forced to find passage through them.

The air which enters into the lower chamber is driven in by a bellows, and after following the passages above described along he surface of the liquors, rises through the apertures or perforations, and bearing the vapor along cómmingles it or them, giving immediate opportunity of that elective attraction, by which they become explosive, and would become so without this agitation in some length of time spontaneously. But by this means they are immediately prepared for use, and in an as much smaller vessel or evaporator than would otherwise be requisite for the reservoir.

It has been before this discovery known, that the vapor of ether would under some circumstances explode. And it is also known, that hydrogen gas will explode when intermixed with oxygen. But the discovery appears to be new to chemists of the highest reputation, that the vapors of alcohol, spirits of turpentine and water will so combine at a temperature about blood heat as to be explosive on the touch of fire. And at a very little lower temperature 70°, they do not evaporate freely enough to give energy to the explosion, or to explode quicky and effectually. So on the contrary at a higher temperature, the vapor rather burns than explodes.

Consequently to govern the temperature I make a second pipe leading from the bellows to the upper part of the evaporator to admit a stream of cooler air.

It does not appear essential in what proportions to each other the liquors are put into the evaporator as the laws of elective attraction among the particles appears to be undisturbed by the quantities, but it is remarked that when water and spirit of turpentine are put in together in equal quantities, or unequal, the air takes up equal parts of each. If alcohol be added to them, the air takes up two parts of the spirit, two of water and one of the turpentine. If alcohol and spirits of turpentine are put in without water, the air takes up nearly eight parts of the former to one of the latter. The water appears therefore to be an important or essential ingredient. At the moment of ignition if unchanged but in form it is decomposed yielding its oxygen and hydrogen incidentally, and momentarily, as the consequence of the explosion forming anew into water. A little steam is therefore produced in the cylinder which would impair the vacuum if not condensed either spontaneously, or by the water allowed to enter the cylinder to keep the lower valve cool.

The evaporator above described is warmed to 80° of Fahrenheit by means of a lamp, or other small quantity of fire, or by the sun's rays, or other source of heat, as fermenting grains, etc.

The vapor thus formed or produced is blown and conducted from the evaporating box or vessel to the cylinder of the engine by means of a bellows. The manner of conducting the vapor which I have found convenient, is first by an upright pipe, which at any suitable height turns horizontally to form the seat of a poppet valve, through which the vapor passes at the proper moment into a horizontal pipe leading into the cylinder a little below the piston, when nearly up; but this part of the pipe is contrived to spread and allow the vapor to flow in moderately by several apertures, that it mix not with the common air which being the heaviest it will not do, but remain floating upon it next the piston. It will be found sufficient to occupy one tenth or one eighth of the cylinder with the vapor.

In order to fire and explode the charge of vapor in the cylinder, there is an opening outward on the pipe leading thereto between it and the valve before mentioned. The opening is closed with a poppet valve and is opened at the instant the other is shut, whereupon a portion of the vapor issues and comes in contact with the flame of a lamp, or any other means of ignition, the fire instantly follows the train to the charge in the cylinder, and the explosion takes place with force enough to expel the air through a large valve at or near the bottom of the cylinder, which is then instantly allowed to close, and the pressure of the atmosphere upon the piston causes it to descend with force or power equal to its area, as in the atmospheric steam engine. The evaporator being analogous to the boiler, and the vapor doing the office of the steam, and the chemical affinities among the particles of the vapor, when ignited doing the office of fuel. It is therefore an improvement on, or going beyond the steam engine, in as much as it does not require a boiler nor the consumption of a body of fuel, or other expensive means of attaining the power of a vacuum, but its elements are in a very simple, original and concentrated form. They are the elements of a new power, expressed by the terms vapor power, or engine.

Having thus described the discovery of the properties of these vapors, the means of producing them in a rapid and prepared manner and the way to throw them into the cylinder of the engine, and the peculiarity of its construction in the simplest form, it remains to describe and explain some of the minute circumstances, and practicable modifications of the machine.

The obstructing valve and the igniting valve in the pipe leading from the evaporator to the cylinder may be moved in succession by any usual mechanical movements. I have used several different movements, but that which is most convenient at present is one taken from the bar which connects the beam with the crank of the flywheel. From it a rod proceeds and accompanies its motion so as to move a small shaft which has starts and catches connected with it to lift and hold in succession these two valves which may be called the charging and the firing valves.

The large valve at the bottom of the cylinder is placed on a seat a little without the cylinder connecting freely with it, and the valve is large in order to allow the expelled air to escape instantly and without essential resistance and noise. This valve is made airtight by means of a leather or otherwise. And that the heat generated at the explosion may not affect it, and for another purpose a small tube is inserted in the opposite side of the passage the outer end of which will be in a vessel of water, so that the instant the vacuum is formed, a jet of cold water will strike the underside of the valve, and at the same time condense any steam that may exist in the cylinder to vitiate the vacuum. This water runs out by another pipe having a valve opening outward while the cylinder is refilling with air.

In order that the cylinder may refill with air each time, the air valve must be opened when the piston arrives down, and must remain open till it has arrived up, and the explosion takes place. To effect these movements I plan a lever or spring whose support is about midway between the center of the cylinder and the center of the valve, and when the piston reaches the end of its stroke it will press upon the lever and start up and raise the valve high enough to be detained by a catch, which works into a notch on its stem. The piston then ascends and the air follows through the open valve. But that the valve may shut the moment the explosion takes place, the catch which suspends the valve must let go, among the mechanical means thereof the following is very simple. Let the catch be connected with an arm, and supported at the angle between them on a pin so that the arm shall extend down to the current of air issuing from the valve, and putting a broad piece of metal or wood upon it to be acted upon by the blast, it will be driven off and unlatch the valve at the instant required.

Although the following auxiliary is not essential, I have found it convenient, as a means of preventing the charge of vapor from mingling with the common air in the cylinder. It may be called a piston because it has the shape of one: and as it is an auxiliary to the effect, it may be called the auxiliary piston: it is made of metal, its place is below the real piston, the rod of which will have a hollow to receive the rod of the other. In operation, this auxiliary will be at first as much below the piston as to measure the space to be occupied by the vapor but not touching the sides of the cylinder. When the vapor enters, the air retires below the auxiliary piston, whose stem is detained in the hollow of the main rod by a spring with just force enough to sustain its weight. On the explosions taking place it acts above or on the upper side of the auxiliary piston and drives it to the bottom of the cylinder, and the air before it, more effectually than otherwise. The piston then decends, and the rod or stem of the auxiliary reenters the bore of the piston rod to its spring which detains and lifts it and carries it to its former place as the piston ascends.

The piston itself it is found convenient to pack with leather notwithstanding there is a momentary ignition below it, but this ignition does not appear to heat the cylinder as the cool air returns into it following the piston both up and down continually.

The piston is more perfectly tight made with leather, in the following manner. First make the bottom of metal, and fit a ring or turn it upward like a dish, so as nearly to fit the cylinder filling the space nearly with a wooden piece firmly securing it to the sheet metal bottom of the piston, surround this with one, two or more rings of leather which will rise above the ring of metal enough to apply closely to the bore of the cylinder. Place around the piston rod a spiral spring or springs, or other elastic substance gently to press the leathers to the bore, and let water or oil be put on the piston to keep the leather moist; it is obvious that the pressure of the atmosphere will cause the leathers to pry more firmly and tight as the piston goes down, but as it returns they must be only gently pressed by the elasticity of the springs. And as it is well known that a flame cannot pass through small openings, so it cannot reach the leather by the small space left between the metal ring and the bore of the cylinder.

So likewise as a flame cannot pass a small aperture, the precaution may be taken to put a wire gauze, across the pipe leading from the evaporator, notwithstanding the stopvalve cuts off the communication, but there could be no other danger even from the explosion of the vapor in the evaporating vessel than the loss of its top. In fact it is not a reservoir of explosive vapor, but an instrument of its formation. It is made at the moment it is required for use, and for this reason the engine to be set in motion will require that the bellows be worked by hand a few moments.

The explosive vapor being lighter than common air, it does not permit of the double-stroke as in the steam engine, but by means of a second cylinder the same effect is produced by an alternate action. And by employing a third cylinder it will be a continued unremitting power, if the three are connected with one shaft; having two cranks one of which is at right angles to the other, indeed four cylinders thus arranged would have the same consequence more powerfully.

But there is still another modification of the engine which seems to have its advantages, and increases the power of one cylinder. Let the cylinder be closed airtight at top, as well as bottom, let the air valve be in or through the piston, and let the cylinder be so prolonged upward as to afford an air chamber above the piston when up. The explosion drives the air from below to the air chamber above the piston, and compresses what was below with what was above into an elastic force, which cooperates with the vacuum formed below. This method requires valves to let in air again below, and let out the compressed air seasonably above.

When the engine is to be applied to a simple operation as that of pumping one cylinder with a flywheel or other balance, to bring up the piston, will assure the purpose.

When for railways, two or three cylinders will be most suitable in order to dispense with a fly or balance having a regular unremitting force.

The power of the vapor engine may be calculated on the same principles as that of the steam engine, viz., multiply the area of the piston in inches, by the weight of the atmosphere 14 3/4 pounds, then by the length of the stroke and the number of strokes per minute and divide by 32,000, the number of pounds which a horse is supposed capable of raising one foot in a minute. An engine whose 2 cylinders should be 1 foot diameter and 3 foot stroke would be a 16 horsepower according to this rule, and require an evaporator of about three feet by two, six or ten inches deep. How long one gallon of spirits of turpentine, with one gallon of whiskey or other cheap spirit will last, is presumable from the smaller quantities heretofore used and may be estimated at hours, though this must be referred to more experience.

When the object of the engine or power is to raise water only the explosion may be made in one vessel to drive the air into another partly occupied with water, which is driven out and upward by a tube to any height equal only to the pressure or elasticity of the air thus produced. Or reversing this operation, if a vacuum be formed at an elevation of 32 feet in a vessel connected by a pipe with water below it will immediately rise and fill the vacant vessel.

The advantages of the vapor engine are as numerous as the objects of its application on a small or great scale of operation. It may be applied to every purpose that the steam engine is used for, and at less expense both in the machinery and fuel. It dispenses with the boiler which is costly when for an atmospheric, or low-pressure engine, and dangerous when for high-pressure. And substitutes the evaporator, which is comparatively light and small.

It is an engine therefore more applicable to the generality of purposes requiring a cheap prime moving power, and this principle is positively well adapted to the purpose of locomotive engine, as being much lighter than the steam engine for that purpose. The locomotive steam engine used at the Killingworth collieries described in the report of Charles Sylvester, civil engineer, weighs 8 tons, and working with 50 pound in the boiler and 30 in the cylinders is a 21 horsepower. But it will probably be found that an equal power may be had by the vapor engine, not exceeding one ton in weight of machinery.

3. Morey's United States Patent

Manuscript preserved in the Baker Library, Dartmouth College, Hanover, N.H.

The United States of America.
To all to whom these Letters Patent shall come:
Whereas Samuel Morey, a Citizen of the United States,
hath alleged that he has invented a new and useful improvement
in the Gas or Vapour Engine

which improvement he states has not been known or used before his application; hath made oath that he does verily believe that he is the true inventor or discoverer of the said improvement; hath paid into the treasury of the United States the sum of thirty dollars, delivered a receipt for the same, and presented a petition to the Secretary of State, signifying a desire of obtaining an exclusive property in the said improvement, and praying that a patent may be granted for that purpose: These are therefore to grant, according to law, to the said Samuel Morey, his heirs, administrators or assigns, for the term of fourteen years, from the first day of April one thousand eight hundred and twenty-six, the full and exclusive right and liberty of making, constructing, using and vending to others to be used, the said improvement; a description whereof is given in the words of the said Samuel Morey himself, in the schedule hereto annexed, and is made a part of these presents.

In testimony whereof, I have caused these Letters to be made Patent, and the Seal of the United States to be hereunto affixed. Given under my hand, at the City of Washington, this first day of April in the year of our Lord one thousand eight hundred and twenty-six and of the independence of the United States of America the fiftieth.

J.Q. Adams
by the President.
City of Washington, to wit:

H. Clay
Secretary of State.

I do hereby certify, that the foregoing Letters Patent were delivered to me on the first day of April in the year of our Lord one thousand eight hundred and twenty-six to be examined; that I have examined the same, and find them conformable to law: and I do hereby return the same to the Secretary of State, within fifteen days from the date aforesaid, to wit; on this first day of April in the year aforesaid.

W. Wirt
Attorney General of the United States.

The schedule referred to in these Letters Patent, and making part of the same, containing a description in the words of the said Samuel Morey himself of his improvement in the gas or vapor engine.

To all persons to whom these presents shall come, be it known, that I, Samuel Morey of Orford in the County of Grafton, and State of New Hampshire, have discovered or invented a new and useful mode of producing power for mechanical purposes, denominated the explosive vapor or gas power or engine, and is described or specified in the words following to wit: That is to say: This discovery consists of a machine or engine, for mechanically mixing or preparing gases or vapors with atmospheric air in such a manner as will render them highly explosive, and exploding the same, thereby forming a vacuum, by means of which a power is derived applicable to nearly every mechanical purpose.

The preparing part of the machine consists of a metallic or other vessel, box or tubes, so constructed that a stream or current of atmospheric air, may pass freed through it, together with the vapor or gas to be made use of through a space interrupted by short turns or other impediments, the objects of which are to blend, mix, and unite them with each other, by which means they are rendered highly explosive.

This admits of an endless variety of forms. A description of one follows, which is found to be perfectly safe and probably as convenient as any. Make an oblong box of tin, or other metallic plates, divide it horizontally, into five or six compartments, by partitions which extend alternately from one end of the box to within a short distance of the other, so that the air entering the lower part, will be compelled by the partitions, to travel the whole length of the box, through each compartment in its ascent to the top. Divide these compartments by vertical partitions, running the whole length of the box, into about half inch spaces, which have the double advantage, of rapidly by their friction of facilitating the mixture of the air and gas with each other, and of also of preventing violence in the explosion in the box, should one take place. A tube for admitting air and other materials into the lower apartment, and another for letting out the explosive or prepared air from the upper one, each covered with fine wire gauze to prevent explosion, will complete the preparing vessel.

The exploding part of the machine or engine also admits of a very great variety of constructions. The following is a convenient one. Have a cylinder fitted with a piston, and connected with a crankshaft as in the steam engine, let the lower end of the cylinder have a valve, of at least one half its diameter, opening outwards. This valve may be made of thin soft leather of the same diameter of the working cylinder, and may be tied or fastened to the lower end of the cylinder, or to the lower end of a condenser attached to the lower end of the cylinder, so as to form a continuation of the same. The lower part of the leather cylinder or valve is to be flattened so as to bring its inner sides together for about four or five inches of its length, and kept in that position by light springs, attached to the two edges of the flat part, placing it in a position much like that of a bow and bowstring. This valve is supported or prevented from being driven into the cylinder by a plate of metal, of sufficient thickness to bear the pressure of the atmosphere, arched or raised outwards and perforated with as many holes of about 1/8 to 1/4 of an inch diameter as can well be made in it. The end of the cylinder or condenser forms the abutments to this arched plate. An air valve also, opening outwards is fired to the side of the cylinder just below the piston when down. A pipe from the preparing box is inserted or attached to the side of the cylinder at from 1/8 to

1/4 of the length of the stroke from the top. This pipe should be as short as possible, when vapor is used to prevent its condensing before it reaches the cylinder, and should be furnished with a valve next the preparing vessel, to cut off the communications with it, and a small valve about half an inch diameter, next the cylinder to let the flame of a lamp communicate with the inflammable air so as to have it take fire and communicate by the trail to the charge in the cylinder. These valves also open upwards, but as the explosion meets with so little resistance below they are never thrown up. They are worked by cams on the crankshaft, as is also the air valve, or by any other convenient mode. To give a double-stroke, it will require two cylinders fitted up in this way communicating with the same preparing vessel and crankshaft. When vapor is intended to be used, put a little alcohol, or high-proof spirit either alone or with the addition of a small part of spirit of turpentine, or put spirit of turpentine and water, or other materials, conveniently capable of evaporation, and the vapor of which is inflammable when mixed with atmospheric air into the lower compartment of the preparing vessel, the proportion of each is not very material, apply a small lamp, or other heating substance under it, to raise the temperature to about blood heat, and place the flame of a lamp at the inflaming valve; care should be taken that the temperature of the box when in use is not much above blood heat, if so it ceases to be explosive from too great proportion of vapor coming over. The piston being down, in its ascent would form a vacuum under it, this is prevented by opening the air valve, which supplies the cylinder with air until the piston reaches the pipe from the preparing box; it then closes and the vapor valve supplies the remainder of the cylinder through the preparing vessel with explosive air, and just before the piston is up, say about 1/8 of an inch, the vapor valve closes and at the same time the inflaming valve opens and shuts, the piston being then still rising, draws or turns the flame of the lamp at the firing valve through the opening into the pipe, inflames the trail and communicates it instantly to the charge in the cylinder, the explosion that ensues, drives out the air from within the cylinder, through the perforated arched plate and leather valve at bottom, which valve instantly collapses, and prevents the return of the air, the steam formed by the explosion and formation of the vacuum, is condensed by keeping the lower part of the cylinder cool and condenser by surrounding them with water, and suffering the vacuum to inject a small stream of cold water near the bottom, which also keeps the arched plate and leather valve cool. The vacuum, under the piston instantly follows the explosion, which descends by the pressure of the atmosphere, and carries the crank with it, while the same process is repeated in the second cylinder, and the power is taken from the crankshaft or piston rod as in the steam engine. When the temperature of the box is low, smaller charges will produce the same effect if a thin metallic plate of the same diameter of the piston be introduced into the cylinder called the charging piston. It is fitted with a small rod which moves through a stuffing box in the main piston, so tight that the friction will support its weight. This charging piston is prevented rising higher than the vapor pipe, by its end striking against a stationary point in its ascent, and forms a partition between the compound air in the cylinder and the common atmospheric air thereby preventing the mixture of so great a proportion of atmospheric air in the cylinder as to lessen the effect of the explosion. When the temperature of the box is such as to give off too much vapor the engine works better by the stopping of the working at all of this charging piston.

This piston should be fully perforated with holes about 1/8 of an inch diameter, lest by accident it might not rise with the working piston, as well as to let the explosion pass freely through it to clear the cylinder. Wire gauze should also be placed between the vapor and inflaming valve, to prevent explosion in the box; should the vapor valve not close in time. When hydrogen gas is intended to be used, an apparatus on the principle of Professor Hare's compound blowpipe for regulating the proportion of air and gas, may be attached to the engine, to throw the air and gas into the preparing box. A box of the form before mentioned, fourteen inches long and four and a half wide, and seven inches deep, will be sufficient with a small lamp, to prepare air fast enough, to furnish from fifty to one hundred charges per minute, for a cylinder, seven or eight inches diameter, having a two foot stroke. The box being in use too only about 1/4 part of the time, it of course would supply in the same way four such cylinders if the air were blown or drawn constantly through it. To keep up the temperature of the box in that case would require the application of more heat, although it appears that the more rapid the current of air is made to pass over the surface of the liquor, the more rapidly it takes up vapor at the same temperature. The following are some of the modes by which I have produced a power from this source.

I have caused common air by the effect of the explosion to be compressed over a volume of water to throw it to a great height or distance.

I have in a measure reversed it, and by forming a vacuum in a vessel above, the water would be driven up by the pressure of the atmosphere.

I have caused the explosion to compress in a reservoir, a quantity of atmospheric air, and make use of that compressed air, for working an engine, similar to those of a double-stroke high pressure steam engine. This mode makes it perfectly safe on account of fire as the compressed air may be led in tubes any distance before it works the engine.

Sometimes I have made a valve in the piston to open upwards, and fill the cylinder below the piston with the prepared air, and when the piston is about halfway up the cylinder, it is at the height of its working stroke, the explosion then takes place, the effect is that the quantity of air above the cylinder is nearly doubled its elasticity or force is also greatly increased by a great increase of its temperature, it now reacts on the piston while the vacuum below adds much to the effect. This mode acts with great energy in a small space.

I have attached to the working piston, thin tubes open at top, about 3/8 of an inch diameter, each one of which is directly over and enters one of the tubes of the condenser, attached to the bottom of the cylinder, which tubes were about 3/4 of an inch diameter the prepared air is let into the cylinder near the upper end as the piston rises up, it fills with the prepared air the upper part of these tubes and the spaces around them, when the piston is about up, the explosion takes place, these tubes prevent all violence in the explosion.

At or near all of those springs which are constantly giving off hydrogen gas, engines may be erected substituting the gas in lieu of the vapor to work constantly for every desired mechanical purpose. The engines can be placed in the most desirable situation, only conducting the hydrogen gas through tubes, and if desired the air and gas may pass in due proportion through the tubes together which with a few short turns, will ensure its suitable preparation, although but a short distance. These explosive vapor engines will work without fire at all when the temperature of the weather in the sun or in the shade is about 80 or 90° if the charges are inflamed by the electric spark. With the addition of a due proportion of ether, according to the temperature of the weather, it is likely the engine will work any time with only a lamp to fire the charges. The less proportion of alcohol is used something higher temperature is required. So again, the more turns the air makes in contact with the liquors in the box, the lower temperature will be required to prepare it. If the temperature of the box is as low as about 70°, the spirit of turpentine refuses to come over at all, or at least in sufficient quantity to give energy to the explosion. I have sometimes attached 30 or 40 tubes about 3/4 of an inch diameter to the bottom of the cylinder, about 8 inches, and kept them surrounded with cold water, the more instantly and perfectly to perfect the vacuum. To cause the piston to move airtight, when the cylinders are made of sheet brass, or copper, I put round the piston a hoop of thin metal from one to five or six inches deep, and a cylinder of leather within that projects above two or more inches, gently pressed to the sides of the cylinder as a flame cannot pass a small space, it will not reach the leather. If the engine be required to work so rapid as to have the heat injure the leather this hoop may be as deep as the length of the stroke, and the stuffing applied at the top of the cylinder to the outside of the hoop, or by lengthening the cylinder, the leather may run under water, oil or whiskey, or canvas may be substituted for leather made tight by gum elastic or other substance that will bear the heat without injury. A small bellows is convenient in putting the engine in motion as the explosive air may be blown into the cylinder to warm it and put it in operation. But as these machines admit in their construction of innumerable modifications, it is impossible to give a description of the whole, that they will even readily admit of, but the objects and discoveries, which are intended to be secured by Patent are few and easily designated, viz: They consist of mixing by mechanical means, in a small compass, atmospheric air and vapor or gas, in a short time and in such manner as will render them highly explosive previous to their entering into the exploding chamber and of making use of given or measured charges of that explosive air and thus producing in that way by explosion a much larger or more extensive vacuum than the space the prepared air previously occupied thereby furnishing in a new way and on new principles a first moving power and the application of that power to useful mechanical operations.

Samuel Morey
Witnesses Alex McIntire William Thornton

4. Press Note on Morey's Vapor Engine

The Franklin Journal and American Mechanics' Magazine. vol. 1 (April 1826) p. 252

Morey's New Vapor Engine

Samuel Morey, Esq. a gentleman whose name is familiar to those who have devoted their attention to mechanical science, has obtained a patent for a vapor engine, which, in the opinion of competent judges, promises to answer well, in practice. The vacuum in the cylinder, is produced, by firing an explosive mixture of atmospheric air, and vapor from common proof spirits, mixed with a small portion of spirits of turpentine. A working model has been set in motion, and kept at work, without elevating the temperature of the fluid, from which the vapor is produced, to a higher degree than that of blood heat. Should no unforeseen difficulties present themselves, in its operation on a large scale, it will be the greatest improvement which has been made for many years, particularly in its application to locomotive engines; as the weight of the materials required, to keep it in action for a considerable length of time, will be so small, as not to be worth mentioning. A gentleman has gone to England, for the purpose of obtaining a patent in that country.

5. Hazard's British Patent "A.D. 1826 ... No. 5402"

Method of Preparing Explosive Mixtures, and Employing them as Motive Power.
Hazard's Specification.

To all to whom these presents shall come, I, Erskine Hazard, a Citizen of the United States of America, but now residing in Norfolk Street, Strand, in the City of Westminster, and County of Middlesex, Engineer, send greeting.

Whereas His present most Excellent Majesty King George the Fourth, by His Letters Patent under the Great Seal of Great Britain, bearing date at Westminster, the Twelfth day of August, One thousand eight hundred and twenty-six, in the seventh year of His reign, did, for Himself, His heirs and successors, give and grant unto me, the said Erskine Hazard, His especial license that I, the said Erskine Hazard, my executors, administrators, and assigns, or such others as I, the said Erskine Hazard, my executors, administrators, or assigns, should at any time agree with, and no others, from time to time and at all times during the term of years therein expressed, should and lawfully might make, use, exercise, and vend, within England, Wales, and the Town of Berwick upon Tweed, and also in all His said Majesty's Colonies and Plantations abroad, the Invention of "A Method or Methods of Preparing Explosive Mixtures, and Employing them as a Moving Power for Machinery," partly communicated by a certain foreigner residing abroad, and additions made by myself; in which said Letters Patent is contained a proviso, obliging me, the said Erskine Hazard, by an instrument in writing under my hand and seal, particularly to describe and ascertain the nature of the said Invention, and in what manner the same is to be performed, and to cause the same to be inrolled in His Majesty's High Court of Chancery within two calender months next and immediately after the date of the said in part recited Letters Patent, as in and by the same, reference being thereunto had, will more fully and at large appear.

Now know ye, that in compliance with the said proviso, I, the said Erskine Hazard, do hereby declare that the nature of the said Invention, and the manner in which the same is to be performed, are particularly to be described and ascertained in and by the Drawings hereunto annexed, and the following description thereof (that is to say):

This Invention is a machine for mechanically preparing mixtures of vapors with atmospheric air, and exploding them, thereby forming a vacuum by means of which a power is derived applicable to mechanical purposes. The preparing part of the machine consists of a metallic or other vessel or tubes so constructed that a stream or current of atmospheric air may pass together with the vapor through a considerable space interrupted by short turns or other impediments, the object of which is to mix and blend them thoroughly with each other, by which they are rendered highly explosive. This admits of an endless variety. The one represented in Figure 1, will probably be found as convenient as any. This is an oblong box of tin or other metallic plates divided horizontally into four compartments by the partitions a, a, a, which extends alternately from one end of the box to within a short distance of the other end, so that air entering the lower compartment at b, would be compelled by the partitions to travel the whole length of the box through each compartment in its ascent to the top of the box. These compartments are divided by the vertical partitions e, e, e,* running the whole length of the box into half inch spaces, which has the double advantage of increasing the friction of the particles of vapor and air with each other, and likewise of preventing violence in an explosion, should one take place, in the box itself. A tube d, for the admission of air and the other materials into the lower compartment, and another e, for the outlet of the explosive mixture from the upper compartment, each covered with a wire gauge f, to prevent explosion, will complete the preparing vessel.

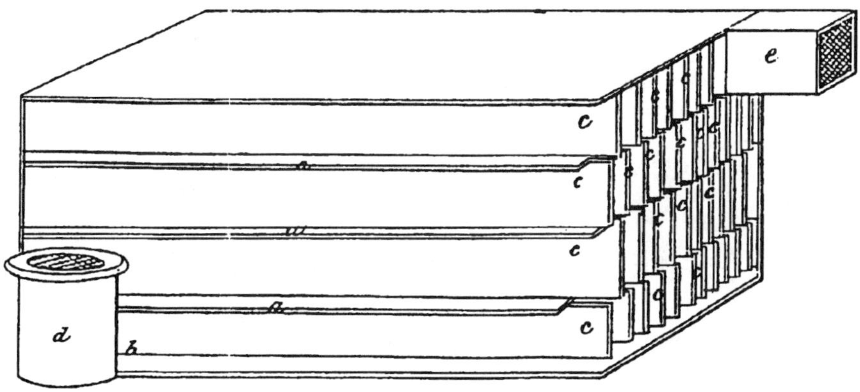

The exploding part of the machine also admits of a great variety of construction; Fig 2, and 3, may serve as an example. A, is a cylinder of brass, copper, or other material fitted with a plunger or piston B, connected with a crankshaft C, as in a steam engine; let the lower end of the cylinder have a valve D, of at least half its diameter opening

[* erratum: read c, c, c,]

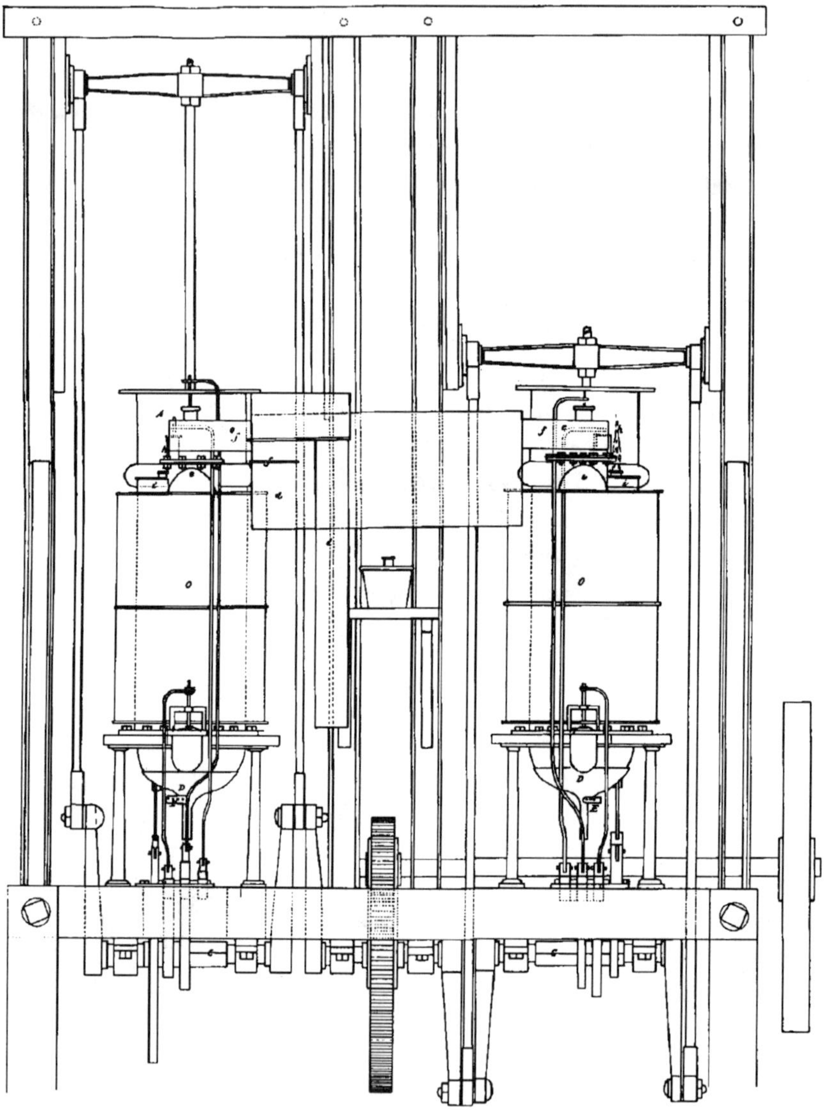

outwards; this valve D, may be made very simply, of a cylinder of leather or other flexible substance of the same diameter as the opening into the cylinder A. This leather is to be tied on the lower end of the cylinder so as to form a continuation of it. The lower end of the leather cylinder is to be flattened so as to bring its sides together for four or five inches of its length, and kept in that position by slight springs E,

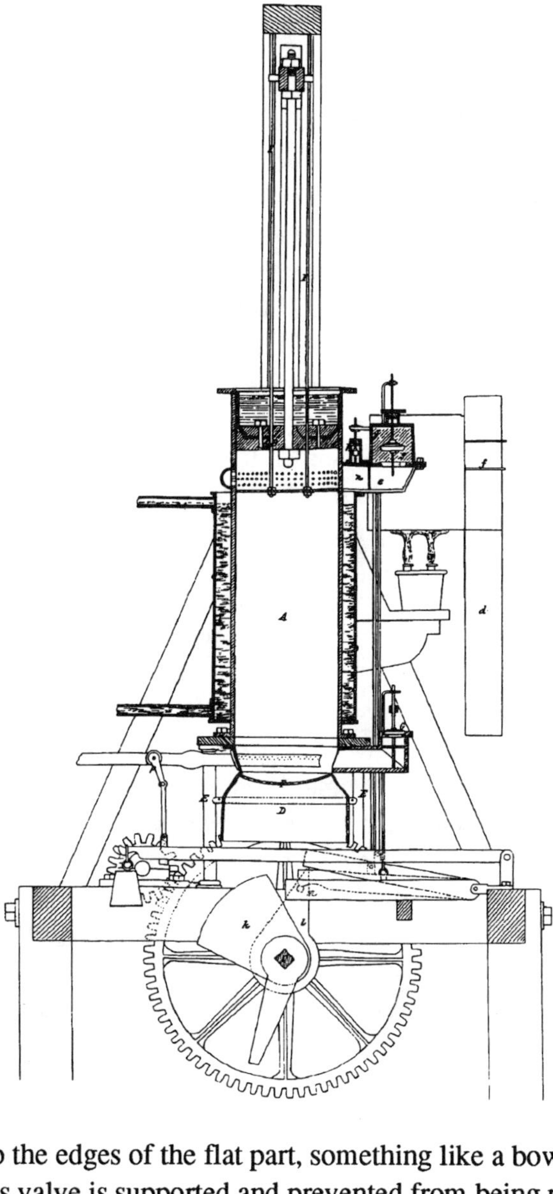

attached to the edges of the flat part, something like a bow and bow-string. This valve is supported and prevented from being pushed into the cylinder by a plate of metal F, of sufficient thickness to bear the pressure of the atmosphere, arched or raised outward, and perforated with as many holes of one eighth or one quarter of an inch diameter as can conveniently be made in it. The end of the cylinder forms the abutment to the arched plate, and air valve G, also opening outward, is fixed in the side of the cylinder just below the plunger or piston when down. The pipe e, from the preparing box is inserted in the side of the cylinder at one sixth to one quarter of the length of the stroke from the top of the cylinder. This pipe or tube should be as short as possible to prevent the vapor from condensing before it reached the cylinder, and should be furnished with a valve g, next to the preparing vessel to cut off the communication with it, and a small valve h, next the cylinder to let the vapor in it communicate with the flame of a lamp i; these valves also open outward, as the explosion meets with so little opposition from the leather valve D, that they are never thrown up. They and the air valve G, are worked by cams k, l, m, on the crankshaft C, or other convenient method. A piston of thin metal plate H, is introduced into the cylinder below the piston B, having a small rod I, worKing sufficiently tight in a stuffing box in the piston B, to raise its weight. This piston H, is prevented from rising with the piston B, higher than the vapor pipe e, by its rod striking a stationary point in its ascent, and thus forms a partition between the explosive mixture and the air in the cylinder. It should be perforated with small holes, lest it by accident should not rise with the piston B, and in that case would prevent the explosion from escaping, and might cause serious accident. A wire gauge n, should also be put between the vapor and inflaming valves g and h, to prevent explosion, should the former not close in time. There are two cylinders fitted up in this way, and communicating with the same preparing vessel and crankshaft. To prepare the explosive mixture put a small quantity of alcohol or alcohol and water, either alone or mixed with spirit of turpentine, or put spirit of turpentine and water, or any other inflammable materials capable of evaporation, into the lower apartment of the preparing vessel, apply a lamp or other heat if necessary under it to raise the temperature and keep the materials about blood warm, and place the flame of a lamp at the inflaming valve. If too much vapor is raised in proportion to the quantity of air the mixture ceases to be explosive. The piston B, being down in its ascent would cause a vacuum; this is prevented by raising the air valve G, which supplies the cylinder with air till B, reaches the pipe e, from the preparing vessel; here the piston H stops, while the piston B continues its motion; the air valve G is closed, and the remainder of the cylinder above H, is supplied with explosive mixture through the pipe e, from the preparing box. Just before the piston B, reaches the end of its stroke, the inflaming valve h opens and shuts, the flame is drawn in, and causes an explosion which drives out the contents of the cylinder through the perforated arched plate F, and flexible valve D, which instantly collapses to prevent the return of the air. The steam formed by the explosion is condensed by keeping the cylinder below the pipe e, cool by surrounding it with water in the jacket o, and by injecting a jet of water through the cock p, (which is also worked by a cam on the crankshaft). The vacuum is formed under the piston B, which descends by the pressure of the atmosphere, and carries the crankshaft with it, while the same process is repeated in the second cylinder; the power is taken from the crankshaft, as in the steam engine. As this machine admits of so much variety of construction by combinations of different pieces common to steam engines and other machinery, I do not intend to claim any particular construction or form, my object being to secure, first, the method of mixing mechanically or by machinery with the inflammable materials, previous to their entering the exploding chamber, such portions of atmospheric air as they would combine with in combustion, thereby rendering them highly explosive, and enabling you to make use of them in an airtight chamber. Secondly, keeping the explosive mixture separate from the atmospheric air in the cylinder and driving out the latter without making use of it in the combustion. Thirdly, the peculiar construction of the flexible valve. And lastly, preparing the explosive mixtures at a temperature much below that of the boiling point.

In witness whereof, I, the said Erskine Hazard, have hereunto set my hand and seal, the Fourth day of October, in the year of our Lord One thousand eight hundred and twenty-six.

(L.S.) Erskine Hazard

And be it remembered, that on the Fourth day of October, in the year of our Lord 1826, the aforesaid Erskine Hazard came before our said Lord the King, in His Chancery, and acknowledged the Specification aforesaid, and all and everything therein contained and specified, in form above written. And also the Specification aforesaid was stamped according to the tenor of the Statute made for that purpose.

Inrolled the Twelfth of October, in the year of our Lord One thousand eight hundred and twenty-six.

Wingfield

6. Morey's Letter to Professor Silliman

The Franklin Journal and American Mechanics' Magazine. vol.2 (August 1826) pp. 115-119
American Journal of Science and Arts. vol. 11 (October 1826) pp. 104-110

An Account of a New Explosive Engine, Generating a Power that May be Substituted for that of the Steam Engine.
By Samuel Morey, of Oxford*, New Hampshire.

To Professor Silliman.

Dear Sir. - Having accidentally discovered that the vapor of water and that of spirit of turpentine, when mixed with a very great proportion of atmospheric air, were highly explosive. I have been endeavouring to produce therefrom, a useful mechanical power; and embrace the earliest convenient opportunity to send you an account, and the result of some experiments. I should hope they may, at least in part, find a place in your Journal.

Alcohol may be substituted for water, or added to it in any proportion. The vapor, and atmospheric air, if placed in contact, will in time unite, as hydrogen gas and common air do, and become apparently as highly explosive. But the process by natural evaporation, would be too slow, when the use or demand was very considerable, unless the reservoir was very large, when it would then be inconvenient, expensive and unsafe. It was desirable to prepare or manufacture the article as wanted. In time it was effected. It was also very desirable, (and there was not, to me, any apparent reason why it might not be effected,) to command or control the explosions, as we do those of gunpowder, although they are much more violent. Another most desirable object, was, to unite in the same engine, if it possibly could be done, the effect or force of the explosion, with that of the vacuum, which always accompanies it; and that, without rendering it too complicated, expensive, and unsafe. By pursuing a course very analogous to that adopted for the use of gunpowder, it was in a measure, if not entirely, effected.

[* erratum: read Orford]

The preparing part of the machine consists of a metallic vessel, or tube, so constructed, that a stream or current of atmospheric air, may pass freely through it, together with the vapor or gas to be made use of; both being impelled through a space interrupted by short turns, or other impediments; the object of which is, to blend, mix, or unite them intimately with each other, by which process they are rendered highly explosive. This apparatus admits of an endless variety of forms. A description of one follows, which is found to be perfectly safe, and probably as convenient as any. Make a box of tin plates, four or five inches wide, and about fourteen long, and seven deep. Divide it horizontally, into four or five compartments, by partitions, which extend from one end of the box, to within a short distance of the other end, so that the air, entering the lower part, will be compelled, by the partitions, to travel the whole length of the box, through each compartment, in its ascent to the top; divide these compartments, by vertical partitions, running the whole length of the box, except at the beginning, or where the air enters or passes out, into spaces, about half an inch apart, which have the double advantage, of effecting by their friction, a rapid mixture of the air and vapor with each other, and also of preventing violence in the explosion, should one take place in the box. A short tube for the admission of air and other materials into the lower apartment, and another for letting out the explosive, or prepared air from the upper one, each covered with fine wire gauze, to prevent explosion, will complete the preparing vessel. The opening to this box should be about two inches in diameter.

The exploding part of the machine, also admits of a great variety of construction. The following may serve as a convenient one. Have a cylinder fitted with a piston or plunger, and connected with a crankshaft, as in the steam engine, let the lower end of the cylinder have a valve of at least half its diameter, opening outwards. This valve may be made of thin soft leather, of the same diameter as the cylinder. This leather is to be tied or fastened to the lower end of the working cylinder, so as to form a continuation of the same. The lower end of the leather cylinder or valve, is to be flattened, so as to bring its inner sides together, for about four or five inches of its length, and kept in that position, by light springs attached to the two edges of the flat part, placing it in a position much like that of a bow, and bowstring. This valve is supported, or prevented from being driven into the cylinder, by a plate of metal, of sufficient thickness to bear the pressure of the atmosphere, arched or raised outwards, and perforated with as many holes as can be well made in it; the holes should be from about 1/8 to 1/4 an inch in diameter. The end of the cylinder forms the abutment to this arched plate. An air valve, also opening outward, is fixed in the side of the cylinder, just below the piston when down. A pipe from the preparing vessel, is inserted, or attached to the side of the cylinder, at from 1/8 to 1/4 of the length of the stroke from the top. This pipe should be as short as it can conveniently be, and it should be furnished with a valve next the preparing box, to cut off the communication with it, and a small valve about half an inch in diameter, next the cylinder, to let the inflammable air communicate with the flame of a lamp, so as to take fire, and communicate, by the trail, to the charge in the cylinder. These valves also open upwards, but as the explosion meets with so little resistance below, they are never thrown upwards. They, as is also the air valve, are worked by beams on the crankshaft, or by any other convenient mode. To give a double-stroke, it will require two cylinders fitted up in this way, communicating with the same preparing vessel and crankshaft. When vapor is intended to be used, put a little alcohol, or high-proof spirit, either alone, or with the addition of a small portion of spirit of turpentine, or put in spirit of turpentine and water, or other convenient materials, capable of evaporation, and the vapor of which is inflammable, when mixed with atmospheric air, into the lower apartment of the preparing vessel; the proportion of each is not very material. Apply a small lamp, or other heating substance, to raise the temperature to about blood heat, and place the flame of a lamp, at the inflaming valve. The piston being down, in its ascent would form a vacuum under it; this is prevented by opening the air valve which supplies the cylinder with common air, until the piston reaches the pipe from the preparing vessel. It then closes, and the vapor valve supplies the remainder of the cylinder, through the preparing vessel, with the explosive air, and just before the piston is up, say about 1/8 of an inch, the vapor valve closes at the same time that the inflaming valve opens and shuts, the piston being then still rising, draws or turns the flame of the lamp at the firing valve, through the opening into the pipe, inflames the trail, and it is instantly communicated to the charge in the cylinder; the explosion that ensues, drives out the air from within the cylinder, through the perforated arched plate and leather valve at bottom, which valve instantly collapses and prevents the return of the air. The steam, formed by the explosion and formation of the vacuum, is condensed, by keeping the lower part of the cylinder cool, by surrounding it with water, and suffering the vacuum to inject a small stream of cold water near the bottom, which also keeps the arched plate, and leather valve, cool. A vacuum under the piston, instantly follows the explosion, which descends by the pressure of the atmosphere, and carries the crank with it, while the same process is repeated in the second cylinder, and the power is taken from the crankshaft, or piston rod, as in the steam engine. When the temperature is low, smaller charges will produce the same effect, if a thin metallic plate of the same diameter as the piston, be introduced into the cylinder, called the charging piston. It is fitted with a small rod, which moves through a stuffing box in the main piston, so tight, that the friction will support its weight. This charging piston is prevented from rising higher than the vapor pipe, by its end striking against a stationary point, in its ascent, and forming a partition between the compound air in the cylinder, and the common atmospheric air, thereby preventing so great a mixture of atmospheric air in the cylinder, as to lessen the effect of the explosion. When the temperature of the box is so high as to give off too great a proportion of vapor, the engine works better, by stopping entirely the working of the charging piston, but not with so good economy. This charging piston should be fully perforated with small holes, lest by accident it should not rise with the working piston, as well as to let the explosion pass freely through it, to clear the cylinder. Wire gauze should also be placed between the vapor and inflaming valve, to prevent explosion in the box, should the vapor valve not close in time. When hydrogen gas is intended to be used, an apparatus similar to Professor Hare's compound blowpipe, may be attached to the engine, to throw the air and gas into the preparing box.

A box of the form and size before mentioned, appears to be sufficient to prepare air fast enough, with a small lamp, to furnish from fifty to one hundred charges per minute, for a cylinder of seven or eight inches in diameter, having a two foot stroke, the box being in use only one quarter part of the time; it, of course, would supply four such cylinders if the air was constantly blown or drawn through it. To keep up the temperature of the box, would, in that case, probably require more heat, but it does appear, that the more rapidly the air is made to pass over the liquors, the more rapidly it takes up vapor at the same temperature.

The following, are some of the methods, I have successfully adopted, in producing a power from this same source.

I have caused the air, by the effect of the explosion, to be compressed over a column of water, to such a degree, as to throw it to a great height and distance.

I have, in a measure, reversed it, and by forming a vacuum in a vessel above, the water would be driven up, by the pressure of the atmosphere.

I have caused the explosion to compress, in a reservoir, a quantity of atmospheric air, and made use of that compressed air, for working an engine, similar to a double-stroke high pressure steam engine.

This mode will make it perfectly safe on account of fire, as the compressed air may be led, in tubes, any distance, before it works the engine.

Sometimes I have made a valve in the piston to open upwards, and fill the cylinder below the piston with the prepared air, and when the piston is about halfway up the cylinder, it is at the height of its working stroke; the explosion then takes place. The effect is, that the quantity of air above the piston is nearly doubled; its elasticity or force is also greatly increased, by a great increase of its temperature, it now reacts on the piston, while a vacuum below, adds greatly to the effect. This mode acts with great energy in a small space.

I have attached to the working piston, thin tubes, about 3/8 of an inch in diameter, open at top, each one of which is directly over and enters one of the tubes of a condenser, attached to the bottom of the cylinder, which tubes are 3/4 of an inch in diameter. The prepared air, as usual, is let into the cylinder near the upper end; as the piston rises, it fills with the prepared air, the upper part of the tubes, and the spaces around them, and when the piston is nearly up, the explosion takes effect, but the tubes prevent all violence.

At, or near all of those springs, which are constantly giving off hydrogen gas, engines may be erected (substituting the gas for that of the vapor of the liquors) to work constantly, for every desired mechanical purpose. The engines can be placed in any chosen situation, by only conducting the gas through tubes, and if desired, the air and gas may pass in due proportion through the tubes together, which, with a few short turns, will insure its suitable preparation, although the distance should be short.

The explosive vapor engines will work without any fire, when the temperature of the weather, in the sun or in the shade, is about 80 or 90°, provided the charge be inflamed by the electric spark. With a due proportion of ether, according to the temperature of the weather, it is probable that the engine will work at any time, with only a lamp to inflame the charge.

The less is the proportion of alcohol the higher temperature will be required. So again, the more turns, back and forth, the air makes in contact with the liquor, in the box, the lower temperature will be required to prepare it. If the temperature of the box gets down to 70°, the spirit of turpentine refuses to come over at all, or at least, in sufficient quantity to give energy to the explosion. I have lately substituted a plunger, in lieu of the piston, and attach the stuffing to the top of the cylinder; it works well, and is more convenient. A small bellows is convenient in putting the engine in motion, or the explosive air may be blown into the cylinder, and inflamed, to warm it, and commence the operation.

A very easy mode, to try an experiment, for the purpose of preparing this explosive air mechanically, is to have made a few feet of inch tubes, of common tin plate. These tubes should be turned, once in about a foot, at right angles, and the long part should be filled with small tubes, about 1/4 of an inch in diameter. If air is made to pass through this crooked tube, while it contains a single spoonful of high-proof whiskey, and spirit of turpentine, with a proper temperature, it comes out highly explosive, if the current is forced through by a hand bellows, with ever so much velocity. As we now construct the engine and preparing vessel, it is impossible that an explosion can take place, that will injure anyone. A few drops of these liquors, on a board in the sun, with a tumbler inverted over them, will explode in a short time, if a flame is applied. It will not, I trust, appear surprising, that these improvements are patented.

I am, dear Sir, yours, most respectfully,

Samuel Morey.

PS. I am sensible, that a drawing ought to accompany this paper, but at present, it is out of my power to furnish one.

7. Extract from Jones' "Remarks upon Explosion and Vapor Engines"

The Franklin Journal and American Mechanics' Magazine. vol. 5 (January 1828) pp. 18-19

It is now upwards of three years, since Mr. Samuel Brown's gas engine was introduced to public attention; and from the ingenuity of its construction, and the sanguine expectations of the projector, and some of his friends, it excited an unusual degree of interest. The principle of its action was by no means novel, the production of a partial vacuum by the combination of explosive mixtures of atmospheric and inflammable air, had been long spoken of, as a substitute for that power obtained by steam. In the year 1814, a model of an engine acting upon this principle, was placed in the hands of the Editor, with a view to his instituting some experiments with it. The inflammable gas to supply it, was to be obtained from the charring of wood; and it was imagined, that the charcoal produced, would go far in defraying the expense of working it. The model was rudely made, but appeared to work with considerable power, and the subject would have been pursued, but for the removal of the Editor to Virginia, on which account the model was returned to the inventor, and it is believed that no further effort was made to perfect it, and thus, the time and money which would probably have been uselessly expended upon it, were saved.

Since that period, several attempts have been made to apply the same principle, either by using inflammable gases, or the vapor from fluids which are volatile, and combustible; hitherto, however, these efforts have been unsuccessful, and we have strong doubts, whether under any ordinary circumstances, an engine operated upon by an explosive mixture of gas, or vapor, will ever be constructed, which will successfully compete with the steam engine.

The want of success, which has attended the ingenious and persevering efforts of Mr. Brown, renders it less necessary to urge our objections, than it would have been two or three years ago. There does not appear, in the successive attempts to perfect this engine, any deficiency of genius, perseverance, or funds; and yet, so far, we believe, it must be pronounced to be a failure. One of the main difficulties which has been experienced, has arisen from the large quantity of gas consumed, which has, to an enormous extent, exceeded the calculations originally made; such at least is the information which we have received, and that, from sources entitled to credit.

... We have spoken of this engine as a failure; we ought, however, to state, that such is not the opinion of the inventor: an engine upon Mr. Brown's principle has recently been constructed in this city, by his son, who came from England for that purpose, under the patronage of one of our most respectable fellow citizens. This engine, although not actually completed, is so far so, as to have been set in operation; the question of its value, in the form it has now assumed, will, therefore, soon be settled. Mr. Morey is also here, with the design of completing one of his explosive engines, an account of which we formerly published. This, in its structure, differs considerably from Brown's; it has two cylinders with pistons; these cylinders are alternately filled with air, and vapor from heated spirits of turpentine; this mixture when ignited by a taper, explodes, expelling nearly all the air through an opening in the bottom of the cylinder, which bottom is formed of leather, in such a way that it operates as a large valve, allowing free exit to the heated and expanded air, and closing so as to prevent its return; the pressure of the atmosphere operates upon each piston, alternately, as it would in an ordinary atmospheric engine with two cylinders.

Engines, analogous in principle to the ordinary steam engine, but acted upon by the elastic force of vapor from different volatile liquids, have also been constructed, or proposed; and several patents have been obtained for them. The objections to these, appear to us insuperable; for although there is no theoretical obstacle in the way of their operation, the practical difficulties which present themselves, have hitherto caused, and are likely to insure, the abandonment of all such attempts; whenever we meet with accounts of such in the journals, we expect the same fate to attend them, which we know await those numerous advertised improvements which sin against the established laws of nature; not that we esteem them like absurd; alcohol, ether, quicksilver, or any other volatile fluid, may be used to work an engine; the attempt, therefore, to employ them, is a legitimate object of pursuit, its attainment is within the bounds of possibility, and although it may ultimately fail, the effort may yet reflect great credit on the individual who makes it. ...

8. Morey's Letter to Messrs. Rush and Muhlenberg

The manuscript must be deemed as lost. Reprinted in: Farrell, G.: Captain Samuel Morey - Who Built a Steamboat Fourteen Years before Fulton. Manchester, N.H., 1915, p. 11

Messrs. Rush and Muhlenburg.*

Dear Sir - Perhaps I ought to have written sooner, but with all my exertions, and they have been as important as they ever were with you, I could not perfect, to my mind, the application of the 'new power' to a boat until within two weeks. It will now run as regular as any that are driven with steam, and with very little expense. The boat is about nineteen feet long, 5 1/2 wide and the engine occupies only about eighteen inches of the stem, and sometimes goes between 7 and 8 miles per hour. The same engine may unquestionably; when in better hands, be made to drive one, properly constructed, of twice the capacity, at least ten. Its application to stationary purposes I perfected last winter. Throughout the whole time I have been constantly perfecting the engine.

I expect to leave here in two or three days for home to arrange my business for winter, and if possible to collect some money for you and Mr. Garrett, as well as some for myself, which I could do were there

---

[* erratum: read Muhlenberg]

any in the country, as I have more than $3,000 of salable personal property and good debts. But whether I get any or not, you may expect to see me next month, if I am alive and well as usual. I have the engine already packed up to be put with the mails on board a Packet, as soon as I can get ready to start from home. Have been inclined to think I should send it to Baltimore in the first instance, but shall postpone entirely anything further. When I have the pleasure to see you I hope to learn what will be the best course to be taken.

It remains for me to have the engine applied to a carriage on a railroad, and when that is done, I should think I have done my part. I can but hope and trust the ensuing winter will see the engine well applied to a carriage on a railroad. With sentiments of the greatest esteem and friendship, I am dear Sirs, as ever

Yours
Sept., 1829
Sam Morey.

9. Extract from Reverend Dana's Letter

Manuscript preserved in the Department of Archives and Manuscripts, Louisiana State University, Baton Rouge, La.

Orford, 15th Oct. 1829

... I understand, they will start for the south today; ... I have not seen Mr. Bissell myself, but am informed, he is going on Capt. Morey's account, who will also go with him. I am told, the Capt. is determined to make one more vigorous effort, to sell his patent right for some of his modern inventions, and that if he does not now succeed, he will give the matter up, and return to Orford, to spend his days in quiet. Believe his invention is, to use gas, instead of steam - he has stated that he could obtain from gas, double the power he could from steam, with half the expense. If these facts could be made to appear I should suppose he would readily dispose of his patent rights, for the wonderful improvement. The great difficulty I suspect, will be, to convince men of these important facts. There have been so many taken in by him, in purchasing his former inventions, that doubtless they will now be more on their guard, and refuse to purchase, till they will be satisfied. They pay their money for something of real value. Should Capt. Morey ... spend the winter at Baltimore ...

10. Jones' Mention of Early American Engine Tests

Jones, T.P.: For the application to machinery of the Power resulting from Explosive Compounds; Henry Rodgers, New York, June 29. Journal of the Franklin Institute. vol. 7 (1833) 6, pp. 409-410

...Many essays have been made to employ the explosive force of gunpowder to the propelling of machinery, and after the expenditure of large sums of money, it has been abandoned as hopeless. ... The editor assisted in the performance of a number of experiments of the same kind, upwards of twenty years ago, ...

11. First and Only European Mention of Morey's Name

Jones, T.P.: Explosive Gas-Engine - Henry Rodgers. Mechanics' Magazine. vol 21 (1834) pp. 110-111

... As regards other explosive compounds, which are all embraced in the sweeping claims of the patentee, the records of the patent-offices might have admonished the present applicant of the futility of such a claim. Brown's gas engine in England, and Morey's explosive engine in America, are familiar examples.

12. Ewbank's Eye-witness Report

Ewbank, T.: Descriptive and Historical Account of Hydraulic and Other Machines for Raising Water. New York 1856, p. 473

In 1826, Mr. Morey, of New Hampshire, patented an explosive engine, and soon after exhibited a large working model in this city, (New York) which we took several opportunities to examine. The piston rods of two vertical and open cylinders were connected to the opposite ends of a vibrating beam. The pistons were made of sheet copper, in the form of plungers, about nine inches diameter, and were made to work airtight by means of a strip of oiled listing or cloth tied round the upper ends of the cylinders. This was all the packing. Mr. Morey employed the vapor of spirits of turpentine and common air.

A small tin dish contained the spirits, and the only heat he used was from a common table lamp. By means of a crank and flywheel a rotary movement was obtained, as in the steam engine.

13. Sellers' Eye-witness Report

Manuscript - "Memoirs Book IV', pp. 63-64 - preserved in the Peal-Sellers Collection, Library of the American Philosophical Society, Philadelphia, Pa. Reprinted in: Ferguson, E.S.. Earl Engine Reminiscenses (1815-1840) of George Escol Sellers. Smithsonian Institution, Washington, D.C., 1965, Bulletin 238

I was going to tell of Mr. Morey's efforts to have one of his interesting engines attached to a hydraulion, but in detail it would be too long a story.

I think the old gentleman came from Hartford, Conn., bringing a working model of what he called by so long a name that I cannot recall it, but the substance was that it was an explosive vapor vacuum engine. The machine he brought had two single-acting cylinders about 10 inches in diameter and he used turpentine or any substance that would, at a low temperature, give an inflammable vapor. He claimed that for fire engines, his engine could be put in operation in less than a minute after lighting his lamps under the vaporizer.

He exhibited his model propelling a bateau on the Schuylkill and it was then brought to the fire engine shops. I do not recollect that he had any plans to attach it to the fire engine pump; I think he only proposed to furnish the driving power.

Finding a set of light small wheels, I think those that had been used on the Perkins garden engine, he proposed to demonstrate the practicability of running the engine to a fire without the aid of men with the long ropes. For the purpose he mounted a kind of buckboard carriage and to prevent firing of this board he placed under his vehicle cylinders a pan containing water. When the explosion took place in the cylinder a flame would shoot through a clap valve at the bottom of the cylinder into the water. At every explosion he would get a vacuum about equal to what was got by the doctors in the old fire cupping process. On the upstroke of Morey's engine, atmospher[ic] air and vapor were drawn in, the valves closing before the piston had quite reached the end of its stroke, in doing which it forced a small hole opposite to which a lamp was burning, the flame of which was drawn into the cylinder and its inflammable mixture exploded and the piston descended by atmospheric pressure into the vacuum.

After Mr. Morey had made what he thought satisfactory yard tests, he had the gate onto Market Street opened that he might mount the seat he had provided, guide the machine and run out Market Street which was then a common road not having been paved with cobble stones although the side gutters were paved with brick, but the narrow paved sidewalks were not curbed.

In starting the machine, somehow Mr. Morey when about mounting his seat lost his hold, tripped and fell flat. As he gathered himself up calling "Stop her, stop her," which no one seemed ready or knew how to do, the thing ran across the street, through the gutter, over the sidewalk and turned a somersault into the brick yard where the clay had been cut away several feet lower than the sidewalk, a complete wreck.

This was the end of the vacuum engine.